# THE NAMES OF PLANTS

### D. GLEDHILL

*Senior Lecturer*
*Department of Botany, Bristol University and*
*Curator of Bristol University Botanic Garden*

SECOND EDITION

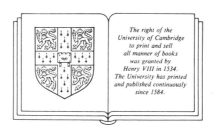

The right of the
University of Cambridge
to print and sell
all manner of books
was granted by
Henry VIII in 1534.
The University has printed
and published continuously
since 1584.

CAMBRIDGE UNIVERSITY PRESS
*Cambridge*
*New York    Port Chester*
*Melbourne    Sydney*

Published by the Press Syndicate of the University of Cambridge
The Pitt Building, Trumpington Street, Cambridge CB2 1RP
40 West 20th Street, New York, NY 10011-4211, USA
10 Stamford Road, Oakleigh, Melbourne 3166, Australia

First published 1985
Second edition 1989
Reprinted 1990

Printed in Great Britain by the University Press, Cambridge

*British Library cataloguing in publication data*
Gledhill, D.
The names of plants – 2nd ed.
1. Plants. Names
I. Title
581'.014

*Library of Congress cataloguing in publication data*
Gledhill, D.
The names of plants / D. Gledhill. – 2nd ed.
p.   cm.
Bibliography: p.
ISBN 0 521 36668 2 – ISBN 0 521 36675 5 (paperback)
1. Botany—Nomenclature.   2. Botany—Dictionaries—Latin.
3. Latin language—Dictionaries—English.   I. Title.
*REF*   QK96.G54   1989
581'.014–dc19   88-34650   CIP
67512
ISBN 0 521 36668 2 hardback
ISBN 0 521 36675 5 paperback

# Contents

# Preface to the first edition

Originally entitled *The naming of plants and the meanings of plant names*, this book is in two parts. The first part has been written as an account of the way in which the naming of plants has changed with time and why the changes were necessary. It has not been my intention to dwell upon the more fascinating aspects of common names but rather to progress from these to the situation which exists today, in which all plant names must conform to internationally agreed standards. I have aimed at producing an interesting text which is equally acceptable to the amateur gardener as to the botanist.

The book had its origins in a collection of Latin plant names and their meanings in English, which continued to grow by the year but which could never be complete. The disparities which I noted between those names which had meaningful translations and those which did not, were added to by the disparities found in some books which not only gave full citation of plant names but also gave a translation of the names, as well as common names, and others which did neither, and by a tendency which may be part of modern language to reduce the names, especially in garden plants, to an abbreviated form (e.g. *Rhodo* for *Rhododendron*). Having myself produced such meaningless common names as Vogel's napoleona and Ivory Coast mahogany by translations of the botanical names (*Napoleonaea vogelii* and *Khaya ivorensis* respectively), I have presented a glossary which should serve to translate the more meaningful, descriptive names of plants from anywhere on earth but which will give little information about many of the people and places epitomized in plant names.

I make no claim that the meanings which I have listed are always the only meanings which have been put upon the various entries. Authors of Latin names have not always explained the meanings of the names which they have erected and as a result such names have subsequently been given different meanings.

## Preface to the second edition

Revision of the text of this book has allowed me to make a number of changes and corrections in both parts. I have attempted to keep the first part acceptable to the amateur gardener by resisting a temptation to make it a definitive guide to the International Code of Botanical Nomenclature. Others have done this already and with great clarity.

Revision has allowed me to include a brief comment on both synonymous and illegitimate botanical names and a reference to recent attempts to accommodate the various traits and interests in the naming and names of cultivated plants.

The revised glossary now includes many commemorative generic names. These do little more than identify the persons for whom the names were raised and their period in history. To give fuller information about each would have made the text very much larger.

# *The nature of the problem*

A rose? By any name?

Man's highly developed constructive curiosity and his
capacity for communication are two of the attributes which
distinguish him from all other animals. Man alone has
sought to understand the whole living world, and things
beyond his own environment, and to pass his knowledge on
to others. Consequently, when he discovers or invents
something new he also creates a new word, or words, in
order to be able to communicate his discovery or invention
to others. There are no rules to govern the manner in which
such new words are formed other than those of their
acceptance and acceptability. This is equally true of the
common, or vulgar, or vernacular names of plants. Such
names present few problems until communication becomes
multilingual and the number of plants named becomes
excessive. For example, the diuretic dandelion is easily
accommodated in European languages as the lion's tooth
(Löwenzahn, dent de lion, dente di leone) or as piss-abed
(Pissenlit, piscacane, piscialetto) but when further study
reveals that there are more than a thousand different kinds
of dandelion throughout Europe then the formulation of
common names for these is both difficult and unacceptable.

Common plant names present language at its richest and
most imaginative (welcome home husband however drunk
you be, for the houseleek or *Sempervivum*; shepherd's
weather-glass, for scarlet pimpernel or *Anagallis*; meet her
i'th'entry kiss her i'th'buttery, or leap up and kiss me, for
*Viola tricolor*; touch me not, for the balsam *Impatiens noli-
tangere*; mind your own business, or mother of thousands,

for *Soleirolia soleirolii*; blood drop emlets, for *Mimulus luteus*). Local variations in common names are numerous and this is perhaps a reflection of the importance of plants in general conversation, in the kitchen and in herbalism throughout the country in bygone days. An often quoted example of the multiplicity of vernacular names is that of *Caltha palustris* for which, in addition to marsh marigold, kingcup and May blobs, there are another ninety local British names (one being dandelion), as well as over 140 German vernacular names and sixty French ones.

Common plant names have many sources. Some came from antiquity by word of mouth as part of language itself, and the passage of time and changing circumstances have obscured their meanings. Fanciful ideas of a plant's association with animals, ailments and festivities, and observations of plant structures, perfumes, colours, habitats and seasonality have all contributed to their naming. So also have their names in other languages. English plant names have come from Arabic, Persian, Greek, Latin, ancient British, Anglo-Saxon, Norman, Low German, Swedish and Danish. Such names were introduced at the same time as the spices, grains, fruit plants and others which merchants and warring nations brought into new areas. Foreign names often remained little altered but some were transliterated in such a way as to lose any meaning which they may have had originally.

The element of fanciful association in vernacular plant names often drew upon comparisons with parts of the body and with bodily functions (priest's pintle for *Arum maculatum*, open arse for *Mespilus germanicus* and arse smart for *Polygonum hydropiper*). Some of these persist but no longer strike us as 'vulgar' because they are 'respectably' modified or the associations themselves are no longer familiar to us (*Arum maculatum* is still known as cuckoo pint (cuckoo pintle) and as wake robin). Such was the sensitivity to indelicate names that Britten and Holland, in their *Dictionary of English Plant Names* (1886), wrote 'We have also purposely excluded a few names which though graphic in their construction and meaning, interesting in their antiquity, and even yet in use in certain counties, are scarcely suited for publication in a work intended for general readers'. They nevertheless included the examples above. The cleaning up of such names was a feature of the Victorian period, during which our common plant names were formalized and reduced in numbers. Some of the resulting names are prissy (bloody cranesbill, for *Geranium sanguineum*, becomes blood-red cranesbill), some are uninspired (naked ladies or meadow saffron, for *Colchicum autumnale*, becomes autumn crocus) and most are not very informative.

This last point is not of any real importance because names do not need to have a meaning or be interpretable. Primarily, names are mere cyphers which are easier to use than lengthy descriptions, and yet, when accepted, they can become quite as meaningful. Within limits, it is possible to use one name for a number of different things, but if the limits are exceeded this may cause great confusion. There are many common plant names which refer to several plants but cause no problem so long as they are used only within their local areas or when they are used to convey only a general idea of the plant's identity. For example, *Wahlenbergia saxicola* in New Zealand, *Phacelia whitlavia* in southern California, USA, *Clitoria ternatea* in West Africa, *Campanula rotundifolia* in Scotland and *Endymion non-scriptus* (formerly *Scilla non-scripta* and now *Hyacinthoides non-scripta*) in England are all commonly called bluebells. In each area, local people will understand others who speak of bluebells, but in all the areas except Scotland the song *The Bluebells of Scotland*, heard perhaps on the radio, will conjure up a wrong impression. At least ten different plants are given the common name of cuckoo-flower in England, signifying only that they flower in spring at a time when the cuckoo is first heard.

The problem of plant names and of plant naming is that common names need not be formed according to any rule and can change as language, or the user of language, dictates. If our awareness extended only to some thousands of 'kinds' of plants we could manage by giving them numbers, but as our awareness extends, more 'kinds' are recognized, and for most purposes we find a need to organize our thoughts about them by giving them names and by forming them into named groups. Then we have to agree with others about the names and the groups, otherwise communication becomes hampered by ambiguity. A completely coded numerical system could be devised but would have little use to the non-specialist without access to the details of encoding.

Formalized names provide a partial solution to the two opposed problems presented by vernacular names: multiple naming of a single plant and multiple application of a single name. The predominantly two-word structure of such formal names has been adopted in recent historic times in all biological nomenclature, especially in the branch which, thanks to Isidorus Hispalensis of Seville (560–636), we now call botany. Of necessity, botanical names have been formulated from former common names but this does not mean that in the translation of botanical names we may expect to find meaningful names in common language. Botanical names, however, do represent a stable system of nomenclature which is usable by people of all nationalities and has relevancy to a system of classification.

Since man became wise, he has domesticated both plants and animals and, for at least the past 300 years, has bred and selected an ever-growing number of 'breeds', 'lines' or 'races' of each, and has given them names. In this, man has accelerated the processes which are believed to be the processes of natural evolution, but has created a different level of artificially sustained and domesticated organisms. The names given by the breeders of the plants of the garden and the crops of agriculture and arboriculture present the same problems as those of vernacular and botanical names. Today, at a time when genetic manipulation promises to afford a way of creating plant races to order, for any feature which is controlled genetically, a new burden of names and additional complications may be produced, adding to the problems already being faced in the search for international agreement over the naming of plants of all kinds, wild, cultivated or items of commerce.

# The size of the problem

'Man by his nature desires to know' (Aristotle)

Three centuries before Christ, Aristotle, disciple of Plato, wrote extensively and systematically of all that was then known of the physical and living world. In this monumental task he laid the foundations of inductive reasoning. When he died, he left his writings and his teaching garden to one of his pupils, Theophrastus (*c.* 370–285 BC), who also took over Aristotle's peripatetic school. Theophrastus' writings on mineralogy and plants totalled 227 treatises, of which nine books of *Historia Plantarum* contain a collection of contemporary knowledge about plants and eight of *De Causis Plantarum* are a collection of his own critical observations, a departure from earlier philosophical approaches, and rightly entitle him to be regarded as the father of botany. These works were subsequently translated into Syrian, to Arabic, to Latin and back to Greek. He recognized the distinctions between monocotyledons and dictoyledons, superior and inferior ovaries in flowers, the necessity for pollination and the sexuality of plants, but although he used names for plants of beauty, use or oddity, he did not try to name everything.

To the ancients, as to the people of earlier civilizations of Persia and China, plants were distinguished on the basis of their culinary, medicinal and decorative uses – as well as their supposed supernatural properties. For this reason plants were given a name and also a description. Theophrastus wrote of some 500 'kinds' of plant which, considering that he had material brought back from Alexander the Great's campaigns, would indicate a considerable lack of discrimination. In Britain we now recognize more than that number of different 'kinds' of moss alone.

Four centuries later, about AD 64, Dioscorides recorded 600 'kinds' of plants and, half a century later still, the elder Pliny, in his huge compilation of the information contained in the writings of 473 authors, described about a thousand 'kinds'. During the 'Dark Ages', despite the remarkable achievements of such people as *Albertus Magnus* (1193–1280) who collected plants during extensive journeys in Europe, and the publication of the *German Herbarius* in 1485 by another collector of European plants, Dr Johann von Cube, little progress was made in the study of plants. It was the renewal of critical observation by Renaissance botanists, such as Dodoens (1517–1585), l'Obel (1538–1616), l'Ecluse (1526–1609) and others, which resulted in the recognition of some 4000 'kinds' of plants by the sixteenth century. At this point in history, the renewal of critical study and the beginning of plant collection throughout the known world produced a requirement for a rational system of grouping plants. Up to the sixteenth century three factors had hindered such classification. The first of these was that the main interested parties were the nobility and apothecaries who conferred on plants great monetary value, either because of their rarity or because of the real or imaginary virtues attributed to them, and regarded them as items to be guarded jealously. Second was the lack of any standardized system of naming plants, and third, and perhaps most important, any expression of the idea that living things could have evolved from earlier extinct ancestors and could therefore form groupings of related 'kinds' was a direct contradiction of the religious dogma of Divine Creation.

Perhaps the greatest disservice to progress was that caused by the doctrine of signatures, which claimed that God had given to each 'kind' of plant some feature which could indicate the uses to which man could put the plant. Thus, plants with kidney-shaped leaves could be used for treating kidney complaints and were grouped together on this basis. Theophrastus Bombast von Hohenheim (1493–1541) had invented properties for many plants under this doctrine and also considered that man had intuitive knowledge of which plants could serve him, and how. He is better known under the Latin name which he assumed, Paracelsus, and the doctrinal book *Dispensatory* is usually attributed to him. The doctrine was also supported by Giambattista Della Porta (1543–1615) who made an interesting extension to it: that the distribution of different 'kinds' of plants had a direct bearing upon the distribution of different kinds of ailment which man suffered in different areas. On this basis, the preference of willows for wet habitats is ordained by God because people who live in wet areas are prone to suffer from rheumatism and, since the bark of *Salix* species gives relief from rheumatic pains (it contains salicylic acid, the analgesic principal of aspirin) the willows are there to serve the needs of man.

In spite of disadvantageous attitudes, renewed critical interest in plants during the sixteenth century led to more discriminating views as to the nature of 'kinds', to searches for new plants from different areas and to concern over the problems of naming plants. John Parkinson (1569–1629), a London apothecary, wrote a horticultural landmark in his *Paradisi in Sole Paradisus Terestris* of 1629. This was an encyclopaedia of gardening and of plants then in cultivation, and contains a lament by Parkinson that in their many catalogues nurserymen 'without consideration of kind or form, or other special note, give(th) names so diversely one from the other, that…very few can tell what they mean'. This attitide towards common names is still with us, but not in so violent a guise as that shown by an unknown writer who, in *Science Gossip* of 1868, wrote that vulgar names of plants presented 'a complete language of meaningless nonsense, almost impossible to retain and certainly worse than useless when remembered – a vast vocabulary of names, many of which signify that which is false, and most of which mean nothing at all.'

Names continued to be formed as phrase-names with a single word name (which was to become the generic name) followed by a description. So we find that the creeping buttercup was known by many names, of which Caspar Bauhin (1550–1664) and Christian Mentzel (1622–1701) listed the following:

Caspar Bauhin, *Pinax Theatri Botanici*, 1623:

*Ranunculus pratensis repens hirsutus* var. C. Bauhin
           *repens fl. luteo simpl.* J. Bauhin
           *repens fol. ex albo variis*
           *repens magnus hirsutus fl. pleno*
           *repens flore pleno*
           *pratensis repens* Parkinson
           *pratensis reptante cauliculo* l'Obel
           *polyanthemos* 1 Dodoens
           *hortensis* 1 Dodoens
           *vinealis* Tabernamontana
           *pratensis etiamque hortensis* Gerard

Christianus Mentzelius, *Index Nominum Plantarum Multi-linguis (Universalis)*, 1682:

*Ranunculus pratensis et arvensis* C. Bauhin
           *rectus acris* var. C. Bauhin
           *rectus fl. simpl. luteo* J. Bauhin
           *rectus fol. pallidioribus hirsutis* J. Bauhin
           *albus fl. simpl. et denso* J. Bauhin
           *pratensis erectus dulcis* C. Bauhin
Ranoncole dolce Italian
Grenoillette dorée o doux Gallic
Sewite Woode Crawe foet English
Suss Hanenfuss
Jaskien sodky Polish
*Chrysanth. simplex* Fuchs

*Ranunculus pratensis repens hirsutus* var. c  C. Bauhin
　　　　　　*repens fl. luteo simpl.*  J. Bauhin
　　　　　　*repens fol. ex albo variis*  Antonius Vallot
　　　　　　*repens magnus hirsut. fl. pleno*  J. B. Tabernamontana
　　　　　　*repens fl. pleno*  J. Bauhin
　　　　　　*arvensis echinatus*  Paulus Ammannus
　　　　　　*prat. rad. verticilli modo rotunda*  C. Bauhin
　　　　　　*tuberosus major*  J. Bauhin
*Crus Galli*  Otto Brunfelsius
*Coronopus parvus Batrachion*  Apuleius Dodonaeus (Dodoens)
*Ranunculus prat. parvus fol. trifido*  C. Bauhin
　　　　　　*arvensis annuus fl. minimo luteo*  Morison
　　　　　　*fasciatus*  Henricus Volgnadius
　　　　　　*Ol. Borrich*  Caspar Bartholino

These were, of course, common or vernacular names with
wide currency and strong candidates for inclusion in lists
which were intended to clarify the complicated state of plant
naming. Local, vulgar names escaped such listing until
much later times, when they were being less used and when
lexicographers began to collect them, saving most from
vanishing for ever.

During the seventeenth century great advances were
made. Robert Morison (1620–1683) published a convenient, or artificial, system of grouping 'kinds' into groups
of increasing size, as a hierarchy. One of his groups we now
call the family Umbelliferae or, to give it its modern name,
Apiaceae, and this was the first natural group to be
recognized. By natural group we imply that the members of
the group share a sufficient number of common features to
suggest that they have all evolved from a common ancestral
stock. Joseph Pitton de Tournefort (1656–1708) had made
a very methodical survey of plants and had assorted 10 000
'kinds' into 698 groups (or genera). The 'kinds' must now
be regarded as the basic units of classification called *species*.
Although critical observation of structural and anatomical
features led to classification advancing beyond the vague

herbal and signature systems, no such advance was made in plant naming until a Swede (of little academic ability when young, we are told) established landmarks in both classification and nomenclature of plants. He was Carl Linnaeus (1707–1778) who classified 7300 species into 1098 genera and gave to each species a binomial name (a name consisting of a generic name-word plus a descriptive epithet, both of Latin form).

It was inevitable that as man grouped the ever-increasing number of known plants (and he was yet only aware of those from Europe and the Mediterranean and a few from other areas), the constancy of associated morphological features in some groups should suggest that the whole was derived, by evolution, from a common ancestor. Morison's family Umbelliferae was a case in point. Also, because the basic unit of any system of classification is the species, and some species were found to be far less constant than others, it was just as inevitable that the nature of the species itself would become a matter of controversy, not least in terms of religious dogma. A point often passed over with insufficient comment is that Linnaeus' endeavours towards a natural system of classification were accompanied by a changing attitude towards Divine Creation. His early view was that the true species were produced by the hand of the Almighty and that abnormal varieties were produced by nature in a sporty mood. In such genera as *Thalictrum* and *Clematis* he concluded that some species were not original creations and, in *Rosa*, he was drawn to conclude that some species had either blended or that one species had given rise to several others. Later, he invoked hybridization as the process by which species could be created, and he attributed to the Almighty the creation of the primeval genera, each with a single species. From the maxims by which he expressed his views in *Genera Plantarum* 6th edn. (Linnaeus, 1764) he attributed the creation of a basic plant, the blending from it of the orders (our families) and from these of the genera to the Creator. Species, he went on, were produced by nature, by hybridization between the genera. The abnormal varieties of the species so formed were the product of chance.

Linnaeus was well aware of the kind of results which plant hybridizers were obtaining in Holland and it is not surprising that his own knowledge of naturally occurring variants led him towards a covertly expressed belief in evolution. However, that expression, and his listing of varieties under their typical species in *Species Plantarum*, where he indicated each with a Greek letter, was still contrary to the dogma of Divine Creation, and it would be another century before an authoritative declaration of evolutionary theory was to be made, by Charles Darwin.

Darwin's essay on *The Origin of Species by Means of Natural Selection* (1859) was published somewhat reluctantly and in the face of fierce opposition. It was concerned with the major evolutionary changes by which species evolve, and was based upon Darwin's own observations on fossils and living creatures. The concept of natural selection, or the survival of any life-form being dependent upon its ability to compete successfully for a place in nature, became, and still is, accepted as the major force directing an inevitable process of organic change. Our conception of the mechanisms and the causative factors for the large evolutionary steps, such as the demise of the dinosaurs and of many plant groups now known only as fossils, and the emergence and diversification of the flowering plants during the last 100 million years, is, at best, hazy.

The great age of plant-hunting, from the second half of the eighteenth century through most of the nineteenth century, produced a flood of species not previously known. Strange and exotic plants were once prized above gold and caused theft, bribery and murder. Trading in 'paper tulips' by the van Bourse family gave rise to the continental stock exchange – the Bourse. With the invention of the Wardian Case by Dr Nathaniel Bagshot Ward, in 1827, it became possible to transport plants from the farthest corners of the world by sea and without enormous losses. The case was a small glasshouse which reduced water losses and made it unnecessary to use large quantities of fresh water on the plants during long sea voyages, and also gave protection from salt spray. In the confusion which was created in naming this flood of plants, and in using many languages to describe them, it became apparent that there was a need for international agreement on both these matters. Today, the rules which have been formulated govern the names of about 300 000 species of plants, which are now generally accepted, and have disposed of a great number of names which were found not to be valid.

Our present state of knowledge about the mechanisms of inheritance and change in plants and animals is almost entirely limited to an understanding of the cause of variation within a species. That understanding is based upon the observed behaviour of inherited characters as first recorded in *Pisum* by Gregor Johann Mendel in 1866. Following the technical development of the microscope, Malpighi (1671), Grew (1672) and others explored the cellular structure of plants and elucidated the mechanisms of fertilization. However, the nature of inheritance and variability remained clouded by myth and monsters until Mendel's work was rediscovered at the beginning of the present century. By 1900, deVries, Correns, Tschermak and Bateson had confirmed that inheritance had a definite, particulate character which is regulated by 'genes'. Sutton (1902) was the first person to clarify the manner in which the characters are transmitted from parents to offspring when he described the behaviour of 'chromosomes' during division of the cell nucleus. Chromosomes are thread-like bodies which can be stained in dividing cells so that the sequence of events of their own division can be followed. Along their length it can be shown that the sites of genetic control, or genes, are situated in an ordered linear sequence. Differences between individuals can now be explained in terms of the different forms, or allelomorphs, in which single genes can exist as a consequence of their mutation.

The concept of a taxonomic species, or grouping of individuals each of which has a close resemblance to the others in every aspect of its morphology, and to which a name can be applied, is not always the most accurate interpretation of the true circumstances in nature. It defines and delimits an entity, but we are constantly discovering that the species is far from being an immutable entity. The botanist may discover that a species has components which have well-defined individual properties (an ability to live on a distinctive soil type, or an adaptation to flower and fruit in harmony with some agricultural practice, or having reproductive barriers caused by differences in chromosome number, etc.) and the horticultural plant breeder may produce a steady stream of new varieties of cultivated species.

If we consider some of the implications of, and attitudes towards, delimiting plant species and their components and naming them, it will become easier to understand the need for internationally accepted rules intended to prevent the unnecessary and unacceptable proliferation of names.

# Towards a solution to the problem

It is basic to the collector's art to arrange items into groups. Postage stamps can be arranged by country of origin and then on face value, year of issue, design, colour variations or defects. The arranging process always resolves into a hierarchic set of groups. In the plant kingdom we have a descending hierarchy of groups through divisions, divided into classes, divided into orders, divided into families, divided into genera, divided into species. Subsidiary groupings are possible at each level of this hierarchy and are employed to rationalize the uniformity of relationships within the particular group. Thus, a genus may be divided into a mini-hierarchy of subgenera, divided into sections, divided into series in order to assort the components into groupings of close relatives. All the components would, nevertheless, be members of the one genus.

Early systems of classification were much less sophisticated and were based upon few aspects of plant structure, such as those which suggested signatures, and mainly upon ancient herbal–medicinal concepts. Later systems would reflect advances in man's comprehension of plant structure and function, and rely more upon the morphology and anatomy of reproductive structures. Groupings such as Natural Orders and Genera had no precise limits or absolute parity one with another, and genera are still very diverse in size, distribution and the extent to which they have been subdivided.

Otto Brunfels (1488–1534) was probably the first person to introduce accurate, objective recording and illustration of plant structure in his *Herbarium* of 1530, and Valerius Cordus (1515–1544) could have revolutionized botany but for his premature death. His four books of German plants contained detailed accounts of the structure of 446 plants, based upon his own systematic studies on them. Many of the plants were new to science. A fifth book on Italian plants was in compilation when he died. Conrad Gesner (1516–1565) published Cordus' work on German plants in 1561 and the fifth book in 1563.

A primitive suggestion of an evolutionary sequence was contained in Matthias de l'Obel's *Plantarum seu Stirpium Historia* (1576) in which narrow-leaved plants, followed by broader-leaved, bulbous and rhizomatous plants, followed by herbaceous dicotyledons, followed by shrubs and trees, was regarded as a series of increasing 'perfection'. Andrea Caesalpino (1519–1603) retained the distinction between woody and herbaceous plants but employed more detail of flower, fruit and seed structure in compiling his classes of plants (*De Plantis*, 1583). His influence extended to the classifications of others who followed him: Caspar Bauhin (1550–1624), who departed from the use of medicinal information and compiled detailed descriptions of the plants to which he gave many binomial names; P. R. de Belleval (1558–1632), who had adopted a naming system consisting of Latin nouns plus Greek adjectival words as binomials; Joachim Jung (1587–1657), whose fear of being accused of heresy prevented him from publishing his work, but some of whose manuscripts survived (these contain many of the terms we still use in describing leaf and flower structure and arrangement, and also contain plant names consisting of a noun qualified by an adjective); Robert Morison (1620–1683), mentioned earlier; John Ray (1627–1705), who introduced the distinction between monocotyledons and dicotyledons but retained the distinction between flowering herbaceous plants and woody plants, and also used binomial names; and Carl Linnaeus who deserves separate mention.

Joseph Pitton de Tournefort (1656–1708) placed great emphasis on the floral corolla and upon defining the genus, rather than the species. His 698 generic descriptions are detailed but his species descriptions are dependent upon binomials and illustrations. Herman Boerhaave (1668–1739) combined the systems of Ray and Tournefort, and others, to incorporate morphological, ecological, leaf, floral and fruiting characters, but none of these early advances received popular support. As Michel Adanson (1727–1806) was to realize, some sixty systems of classification had been proposed by the middle of the eighteenth century and none had been free from narrow conceptual restraints. His plea that attention should be focused on 'natural' classification through processes of inductive reasoning, because of the wide range of characteristics then being employed, did not enjoy wide publication and his work was not well regarded when it did become more widely known.

Before considering the major contributions made by Carl Linnaeus, it should be noted that the names of many higher groups of plants, of families and of genera, were well established at the beginning of the eighteenth century and that several people had used simplifed, binomial names for species. Indeed, August Quirinus Rivinus (1652–1723) had proposed that no plant should have a name of more than two words.

Carl Linnaeus (1707–1778) was the son of a clergyman, Nils, who had adopted the latinized family name when he became a student of theology. Carl also went to theological college for a year but then left and became an assistant gardener in Prof. Olof Rudbeck's botanic garden at Uppsala. His ability as a collector and arranger soon became evident and, after undertaking tours through Lapland and Holland, he began to publish works which are now the starting points for naming plants and animals. In literature he is referred to as Carl or Karl or Carolus Linnaeus, Carl Linné (an abbreviation) and, later in life, as Carl von Linné. His life became one of devotion to the classification and naming of all living things and of teaching others about them. His numerous students played a very important part in the discovery of new plants from many parts of the world. Linnaeus' main contribution to botany was his method of naming plants, in which he combined Bauhin's and Belleval's use of binomials with Tournefort's and Boerhaave's concepts of the genus. His success, where all others before him had failed, was due to his early publication of his most popular work, an artificial system of classifying plants. In this he employed the number, structure and disposition of the stamens of the flower to define twenty-three classes, each subdivided into orders on the basis of the number of parts constituting the pistil, with a twenty-fourth class containing those plants which had their reproductive organs hidden to the eye, the orders of which were the ferns, mosses, algae (in which he placed liverworts, lichens and sponges), fungi and palms. This sexual system provided an easy way of grouping plants and of allocating newly discovered plants to a group. Originally designed to accommodate the plants of his home parish, it was elaborated to include first the Arctic flora and later the more diverse and exotic plants being discovered in the tropics. It continued in popular use into the nineteenth century despite its limitation of grouping together strange bedfellows: red valerian, tamarind, crocus, iris, galingale sedge and mat grass are all grouped under *Triandria* (three stamens) *Monogynia* (pistil with a single style).

In 1735, Linnaeus published *Systema Naturae*, in which he grouped species into genera, genera into orders and orders into classes on the basis of structural similarities. This was an attempt to interpret evolutionary relationships or assemblages of individuals at different levels. In *Species Plantarum*, published in 1753, he gave each species a binomial name. The first word of each binomial was the name of the genus to which the species belonged and the

18

second word was a descriptive, or specific epithet. Both words were in Latin or Latin form. Thus, the creeping buttercup he named as *Ranunculus repens*.

It now required that the systematic classification and the binomial nomenclature which Linnaeus had adopted should become generally accepted and, largely because of the popularity of his sexual system, this was to be the case. Botany could now contend with the rapidly increasing number of species of plants being collected for scientific enquiry, rather than for medicine or exotic gardening as in the seventeenth century. For the proper working of such standardized nomenclature, however, it was necessary that the language of plant names should also be standardized.

Linnaeus' views on the manner of forming plant names, and the use of Latin for these and for the descriptions of plants and their parts, have given rise directly to modern practice and a Latin vocabulary of great versatility, but which would have been largely incomprehensible in ancient Rome. He applied the same methodical principles to the naming of animals, minerals and diseases and, in doing so, established Latin, which was the *lingua franca* of his day, as the internationally used language of science and medicine.

The rules by which we now name plants depend largely on Linnaeus' writings, but for the names of plant families we are much dependent on A. L. de Jussieu's classification in his *Genera Plantarum* of 1789. For the name of a species, the correct name is that which was first published since 1753. This establishes Linnaeus' *Species Plantarum* (associated with his *Genera Plantarum* 5th edn of 1754 and 6th edn of 1764) as the starting point for the names of species (and their descriptions).

Linnaeus' sexual system of classification was very artificial and, although Linnaeus must have been delighted at its popularity, he regarded it as no more than a convenient pigeon-holing system. He published some of his views on grouping plant genera into natural orders (our families) in

*Philosophia Botanica* (1751). Most of his orders were not natural groupings but were considerably mixed assemblages. By contrast, Bernard de Jussieu (1699–1777), followed by his nephew Antoine Laurent de Jussieu (1748–1836), searched for improved ways of arranging and grouping plants as natural groups. In A. L. de Jussieu's *Genera Plantarum* the characteristics are given for 100 plant families, and most of these we still recognize.

Augustin Pyrame de Candolle (1778–1841) also sought a natural system, as did his son Alphonse, and he took the evolutionist view that there is an underlying state of symmetry in the floral structure which we can observe today and that by considering relationships in terms of that symmetry natural alliances may be recognized. This approach resulted in a great deal of monographic work from which de Candolle formed views on the concept of a core of similarity, or type, for any natural group and the need for control in the naming of plants.

Today, technological and scientific advances have made it possible for us to recognize subcellular, chemical and the minutest of morphological features and to incorporate as many items of information as are available about a plant in computer-aided assessments of that plant's relationships to others. Biological information has often been found to conflict with the concept of the taxonomic species, and there are many plant groups in which the 'species' can best be regarded as a collection of highly variable populations. The gleaning of new evidence necessitates a continuing process of reappraisal of families, genera and species. Such reappraisal may result in subdivision or even splitting of a group into several new ones or, the converse process, in lumping together two or more former groups into one new one. Since the bulk of research is carried out on the individual species, most of the revisions are carried out at or below the rank of species. On occasion, therefore, a revision at the family level will require the transfer of whole genera from one family to another, but it is now more common for a revision at the level of the genus to require the transfer of some, if not all, the species from one genus to another. Such revisions are not mischievous but are the necessary process by which newly acquired knowledge is incorporated into a generally accepted framework. It is because we continue to improve the extent of our knowledge of plants that revision of the systems for their classification continues and, consequently, that name changes are inevitable.

The equivalence, certainly in evolutionary terms, of groups of higher rank than of family is a matter of philosophical debate, and even at the family level we find divergence of views as to whether those with few components are equivalent to those with many components. Because the taxonomic species is the basic unit of any system of classification, we have to assume parity between species; that is to say, we assume that a widespread species is in every way comparable with a rare species which may be restricted in its distribution to a very small area.

It is a feature of plants that their diversity of habit, longevity, mode of reproduction and tolerance to environmental conditions presents a wide range of biologically different circumstances. Faced with the taxonomic problem of delimiting, defining and naming a species, we have to identify a grouping of individuals whose characteristics are sufficiently stable to be defined, so that a name can be applied to the group and a 'type', or exemplar, can be specified for that name. It is because of this concept of the 'type' that changes have to be made in names of species in the light of new discoveries, and that entities below the rank of species have to be recognized. Thus, we speak of a botanical 'subspecies' when part of the species grouping can be distinguished as having a number of features which remain constant and as having a distinctive geographical or ecological distribution. When the degree of departure from the typical material is of a lesser order we may employ the inferior category of 'variety'. The term 'form' is employed to describe a variant which is distinct in a minor way only, such as a single feature difference which might appear sporadically due to genetic mutation or sporting.

The patterns and causes of variation differ from one species to another, and this has long been recognized as a problem when trying to reconcile the idea of a taxonomic species with that of a biological system of populations in perpetual evolutionary flux. Below the level of species, agreement about absolute ranking is far from complete, and even the rigidity of the infraspecific hierarchy (subspecies, varietas, subvarietas, forma, sub-forma) is now open to question.

When a new name has to be given to a plant which is widely known under its superseded old name it is always a cause of annoyance. Gardeners always complain about such name changes but there is no novelty in that. On the occasion of Linnaeus being proposed for Fellowship of the Royal Society, Peter Collinson wrote to him in praise of his *Species Plantarum* but at the same time complained that Linnaeus had introduced new names for so many well-known plants.

The gardener has some cause to be aggrieved by changes in botanical names. Few gardeners show much alacrity in adopting new names and perusal of gardening books and catalogues shows that horticulture seldom uses botanical names with all the exactitude which they can provide. Horticulture, however, not only agreed to observe the international rules of botanical nomenclature but also formulated its own additional rules for the naming of plants grown under cultivation. It might appear as though the botanist realizes that he is bound by the rules, whereas the horticulturalist does not, but to understand this we must recognize the different facets of horticulture. The rules are of greatest interest and importance to specialist plant breeders and gardeners with a particular interest in a certain plant group. For the domestic gardener it is the growing of beautiful plants which is the motive force behind his activity. Between the two extremes lies every shade of interest and the main emphasis on names is an emphasis on garden names. Roses, cabbages, carnations and leeks are perfectly adequate names for the majority of gardeners, but if greater precision is needed, the gardener can ask for the name of the variety. Consequently, most gardeners are satisfied with a naming system which has no recourse to the botanical rules whatsoever. Not surprisingly, therefore, seed and plant catalogues also avoid botanical names. The specialist plant breeder, however, shows certain similarities to the apothecaries of an earlier age. Like them he guards his art and his plants jealously because they represent the source of his future income and, also like them, he has the desire to understand every aspect of his plants. The apothecaries gave us the first centres of botanical enquiry and the plant breeders of today give us the new varieties which are needed to satisfy our gardening and food-production requirements. The commercial face of plant breeding, however, attaches a powerful monetary significance to the names given to new varieties.

Botanical names have occasionally to be resorted to by gardeners when they discover some cultural problem with a plant which shares the same common name with several different plants. Lately, the Guernsey lily, around which has always hung a cloud of mystery, has been offered to the public in the form of *Amaryllis belladonna* L. The true Guernsey lily has the name *Nerine sarniensis* Herb. (but was named *Amaryllis sarniensis* by Linnaeus) The epithet *sarniensis* means 'of Sarnia' or 'of Guernsey'. Sarnia was the old name for Guernsey, and is an example of a misapplied geographical epithet, since the plant's native area is South Africa. Some would regard the epithet as indicating the fact that Guernsey was the first place in which the plant was cultivated. This is historically incorrect, however, and does nothing to help the gardener who finds that the Guernsey lily which he has bought does not behave, in culture, as *Nerine sarniensis* is known to behave. This example is one involving a particularly contentious area as to the taxonomic problems of generic boundaries and typification, but there are many others in which common and Latin garden names are used for whole assortments of garden plants, ranging from species (*Nepeta mussinii* and *N. cataria* are both catmint) to members of different genera (japonicas including *Chaenomeles speciosa* and *Kerria japonica*) to members of different families (*Camellia japonica* is also a japonica, as are the last two species above, and the diversity of bluebells was mentioned earlier).

New varieties, be they timber trees, crop plants or garden flowers, require names and those names need to be definitive. As with the earlier confusion of botanical names (different names for the same species or the same name for different species), so there can be the same confusion of horticultural names. As will be seen, rules for cultivated plants require that new names have to be established by publication. This gives to the breeder the commercial advantage of being able to supply to the public his new variety under what, initially, amounts to his mark of copyright. Indeed, in some parts of the world the breeder's new varietal name for a plant actually represents a trade mark and it may not be long before genetically engineered animals and plants can be protected under patents legislation.

# The rules of botanical nomenclature

The rules which now govern the naming and the names of plants really had their beginnings in the views of A. P. de Candolle as he expressed them in his *Théorie Élémentaire de la Botanique* (1813). There, he advised that plants should have names in Latin (or Latin form, but not compounded from different languages), formed according to the rules of Latin grammar and subject to the right of priority for the name given by the discoverer or the first describer. This advice was found to be inadequate and, in 1862, the International Botanical Congress in London adopted control over agreements on nomenclature. Alphonse de Candolle (1806–1893) who was A. P. de Candolle's son, drew up four simple 'Lois', or laws, which were aimed at resolving what threatened to become a chaotic state of plant nomenclature. The Paris International Botanical Congress of 1867 adopted the Lois, which were:

1. One plant species shall have no more than one name.
2. No two plant species shall share the same name.
3. If a plant has two names, the name which is valid shall be that which was the earliest one to be published after 1753.
4. The author's name shall be cited, after the name of the plant, in order to establish the sense in which the name is used and its priority over other names.

It can be seen from these first rules that until then botanists frequently gave names to plants with little regard to either the previous use of the same name or the existing names which had already been applied to the same plant. It is because of this aspect that one often encounters the words *sensu* and *non* inserted before the name of an author, although both terms are more commonly used in the sense of taxonomic revision, and indicate that the name is being used 'in the sense of' or 'not in the sense of' that author, respectively.

The use of Latin as the language in which descriptions and diagnoses were written was not universal in the nineteenth century, and many regional languages were used in different parts of the world. A description is an account of the plant's habit, morphology and periodicity, whereas a diagnosis is an author's definitive statement of the plant's diagnostic features and circumscribes the limits outside which plants do not pertain to that named species. A diagnosis often states particular ways in which the species differs from another species of the same genus. Before the adoption of Latin as the accepted language of botanical nomenclature, searching for names already in existence, and confirming their applicability to a particular plant, involved searching through literature in many languages. The requirement to use Latin was written into the rules by the International Botanical Congress in Vienna in 1905. However, the American Society of Plant Taxonomists produced its own Code in 1947, which became known as the Brittonia edition of the Rules or the Rochester Code, and disregarded this requirement. Not until 1959 was international agreement achieved and then the requirement to use Latin was made retroactive to January 1st, 1935: the year of the Amsterdam meeting of the Congress.

The rules are considered at each International Botanical Congress, formerly held at five-yearly and more recently at six-yearly intervals during peacetime. The International Code of Botanical Nomenclature (first published as such in 1952) was formulated at the Stockholm Congress of 1950. In 1930, the matter of determining the priority of specific epithets was the main point at issue. The practice of British botanists had been to regard the epithet which was first published after the plant had been allocated to its correct genus as the correct name; this has been called the Kew Rule. But it was defeated in favour of the rule which now gives priority to the epithet which was the first to be published from the starting date of May 1st, 1753. Epithets which predate the starting point, but which were adopted by Linnaeus, are attributed to Linnaeus (e.g. Bauhin's *Alsine media*, *Ammi majus*, *Anagyris foetida* and *Galium rubrum* and Dodoens' *Angelica sylvestris* are examples of binomials credited to Linnaeus).

The 1959 International Botanical Congress in Montreal introduced the requirement under the code that for valid publication of a name the author of that name (of a family or any taxon of lower rank) should cite a 'type' for that name, and that this requirement should be retroactive to January 1st, 1958. The idea of a type goes back to A. P. de Candolle and it implies a representative collection of characteristics to which a name applies. The type in Botany is a nomenclatural type: it is the type for the name and the name is permanently attached to it or associated with it. For the name of a family, the representative characteristics which that name implies are those embodied in one of its genera, which is called the type genus. In a similar way, the type for the name of a genus is the type species of that genus. For the name of a species or taxon of lower rank, the type is a specimen lodged in an herbarium or, in certain cases, published illustrations. The type need not, nor could it, be representative of the full range of entities to which the name is applied. Just as a genus, although having the features of its parent family, cannot be fully representative of all the genera belonging to that family, so no single specimen can be representative of the full range of variety found within a species.

For the name to become the correct name of a plant or plant group, it must satisfy two sets of conditions. First, it must be constructed in accordance with the rules of name formation, which ensures its legitimacy. Second, it must be published in such a way as to make it valid. Publication has to be in printed matter which is distributed to the general public or, at least, to botanical institutions with libraries accessible to botanists generally. Since January 1st, 1953, this has excluded publication in newspapers and trade catalogues. Valid publication also requires the name to be accompanied by a description or diagnosis, an indication of its rank and the nomenclatural type, as required by the rules. This publication requirement, and subsequent citation of the new name followed by the name of its author, ensures that a date can be placed upon the name's publication and that it can, therefore, be properly considered in matters of priority.

The present scope of the Code is expressed in the Principles, which have evolved from the de Candollean Lois:

1. Botanical nomenclature is independent of zoological nomenclature. The Code applies equally to names of taxonomic groups treated as plants whether or not these groups were originally so treated.
2. The application of names of taxonomic groups is determined by means of nomenclatural types.
3. The nomenclature of a taxonomic group is based upon priority of publication.
4. Each taxonomic group with a particular circumscription, position and rank can bear only one correct name, the earliest that is in accordance with the rules, except in specified cases.
5. Scientific names of taxonomic groups are treated as Latin regardless of their derivation.
6. The rules of nomenclature are retroactive unless expressly limited.

The detailed rules are contained in the Articles and Recommendations of the Code, and mastery of these can only be gained by practical experience. A most lucid summary and comparison with other Codes of biological nomenclature is that of Jeffrey (1978) written for the Systematics Association.

There are still new species of plants to be discovered and an enormous amount of information yet to be sought for long-familiar species, particularly evidence of a chemical nature, and especially that concerned with proteins, which may provide reliable indications of phylogenetic relationships. For modern systematists, the greatest and most persistent problem is our ignorance about the apparently explosive appearance of a diverse array of flowering plants, some 100 million years ago, from one or more unknown ancestors. Modern systems of classification are still frameworks within which the authors arrange assemblages in sequences or clusters to represent their own idiosyncratic interpretation of the known facts. In addition to having no firm record of the early evolutionary pathways of the flowering plants, the systematist also has the major problems of identifying clear-cut boundaries between groups and of assessing the absolute ranking of groups. It is because of these continuing problems that, although the Code extends to taxa of all ranks, most of the rules are concerned with the names and naming of groups from the rank of family downwards.

Before moving on to the question of plant names at the generic and lower ranks, this is a suitable point at which to comment on the move towards the standardization of the names of families in recent years. The new names for families are now starting to appear in books and catalogues, and some explanation in passing may help to dispel any confusion.

## *Family names*

Each family can have only one correct name, and that of course is the earliest legitimate one, except in cases of limitation of priority by conservation. In other words, there is provision in the Code for disregarding the requirement of priority when a special case is proved for a name to be conserved. Conservation of names is intended to avoid disadvantageous name changes, even though the name in question does not meet all the requirements of the Code. Names which have long-standing use and wide acceptability and are used in standard works of literature can be proposed for conservation, and when accepted need not be discarded in favour of new and more correct names.

The names of families are plural adjectives used as nouns and are formed by adding the suffix *-aceae* to the stem, which is the name of an included genus. Thus, the buttercup genus *Ranunculus* gives us the name Ranunculaceae for the buttercup family, and the water-lily genus *Nymphaea* gives us the name Nymphaeaceae for the water-lilies. A few family names are conserved, for the reasons given above, which do have generic names as their stem, although one, the Ebenaceae, has the name *Ebenus* Kuntze (1891) *non* Linnaeus (1753) as its stem. Kuntze's genus is now called *Maba* but its parent family retains the name Ebenaceae even though *Ebenus* L. is the name used for a genus of the pea family. There are eight families for which specific exceptions are provided and which can be referred to either by their long-standing conserved names or, as is increasingly the case in recent floras and other published works on plants, by their names which are in agreement with the Code. These families, and their equivalent names are:

Compositae      or Asteraceae (on the genus *Aster*)
Cruciferae       or Brassicaceae (on the genus *Brassica*)
Gramineae       or Poaceae (on the genus *Poa*)
Guttiferae       or Clusiaceae (on the genus *Clusia*)
Labiatae          or Lamiaceae (on the genus *Lamium*)
Leguminosae   or Fabaceae (on the genus *Faba*)
Palmae           or Arecaceae (on the genus *Areca*)
Umbelliferae    or Apiaceae (on the genus *Apium*)

Some botanists regard the Leguminosae as including three

30

sub-families, but others accept those three components as each having family status. In the latter case, the three families are the Caesalpiniaceae, the Mimosaceae and the Papilionaceae. The last of these family names refers to the resemblance of the pea- or bean-flower structure, because of its large and colourful sail petal, to a resting butterfly (Papilionoidea) and is not based upon the name of a plant genus. If a botanist wishes to retain the three-family concept, the name Papilionaceae is conserved against Leguminosae and the modern equivalent is Fabaceae. Consequently, the Fabaceae are either the entire aggregation of leguminous plant genera or that part of the aggregate which does not belong in either the Caesalpiniaceae or the Mimosaceae.

Some eastern European publications use Daucaceae for the Apiaceae and split the Asteraceae into Carduaceae and Chicoriaceae, and also adopt various views as to the generic basis of family names (e.g. Oenotheraceae for Onagraceae by insisting that Linnaeus' genus *Oenothera* has prior claim over Miller's genus *Onagra*).

## Generic names

The name of a genus is a noun, or word treated as such, and begins with a capital letter. It is singular and may be taken from any source whatever and may even be composed in an absolutely arbitrary manner. The etymology of generic names is, therefore, not always complete and, even though the derivation of some may be discovered, they lack meaning. By way of examples:

> *Portulaca* from the Latin *porto* (I carry) and *lac* (milk) translates as 'Milk-carrier'.

> *Pittosporum* from the Greek πιττόω (I tar) and σπόρος (a seed) translates as 'Tar-seed'.

> *Hebe* was the goddess of youth and, amongst other things, the daughter of Jupiter. It cannot be translated further.

> *Petunia* is taken from the Brazilian name for tobacco.

*Tecoma* is taken from a Mexican name.

*Linnaea* is one of the names which commemorate Linnaeus.

*Sibara* is an anagram of *Arabis*.

*Aa* is the name which Reichenbach gave to an orchid genus, which he separated from *Altensteinia*. It has no meaning and, as others have observed, must always appear first, in an alphabetic listing.

If all specific names were constructed in an arbitrary manner there would have been no enquiries of the writer and this book would not have been written. In fact, the etymology of plant names is a rich store of historical interest and conceals many facets of humanity, ranging from the sarcasm of some authors to the humour of others. This is made possible by the wide scope available to authors for formulating names, and because whatever language is the source, names are treated as being in Latin. Imaginative association has produced some names which are very descriptive, provided that the reader can spot the association. In the algae, the chrysophyte that twirls like a ballerina has been named *Pavlova gyrans*, and in the fungi, a saprophyte on leaves of *Eucalyptus* which has a wide mouthed spore-producing structure has been named *Satchmopsis brasiliensis* (Satchmo, satchelmouth). The large vocabulary of botanical Latin comes mostly from the Greek and Latin of ancient times, but since the ancients had few words which related specifically to plants and their parts, a Latin dictionary is of somewhat limited use in trying to decipher plant diagnoses. By way of examples, Table 1 gives the parts of the flower (Latin *flos*, Greek ἄνθος) (illustrated in Fig. 1) and the classical words from which they are derived, together with their original sense.

The grammar of botanical Latin is very formal and much more simple than that of the classical language itself. A full and most authoritative work on the subject is contained in Stearn's book *Botanical Latin* (1983). Nevertheless, it is

Table 1

| Flower part | Greek | Latin | Former meaning |
|---|---|---|---|
| calyx | κάλυξ κύλιξ | — — | various kinds of covering cup or goblet |
| sepal | σκέπη | — | covering |
| corolla | — | corolla | garland or coronet |
| petal | πέταλον | — | leaf |
| stamen | — | petalum | metal plate |
| filament | — | stamen | thread, warp, string |
| anther | — | filamentum | thread |
| androecium | ἀνδρ-, οἰκός | anthera | potion of herbs |
| stigma | στίγμα | — | man, house |
| style | στῦλος | — | tattoo or spot |
| carpel | καρπός | stilus | pillar or post |
| gynoecium | γυνή, οἰκός | — | pointed writing tool |
| pistil | — | — | fruit |
|  |  | pistillum | woman, house |
|  |  |  | pestle |

helpful to know that Latin nouns (such as family and generic names) have gender, number and case and that the words which give some attribute to a noun (as in adjectival specific epithets) must agree with the noun in each of these. Having gender means that all things (the names of which are called nouns) are either masculine or feminine or neuter. In English we treat almost everything as neuter, referring to nouns as 'it' except animals and most ships and aeroplanes (which are commonly held to be feminine). Gender is explained further below. Number means that things may be single (singular) or multiple (plural). In English we either have different words for the singular and plural (man and men, mouse and mice) or we convert the singular into the plural most commonly by adding an 's' (ship and ships, rat and rats) or more rarely by adding 'es' (box and boxes, fox and foxes) or rarer still by adding 'en' (ox and oxen). In Latin, the difference is expressed by changes in the endings of the words. Case is less easy to understand but means the significance of the noun to the meaning of the sentence in which it is contained. It is also expressed in the endings of the words. In the sentence 'The flower has charm', the flower is singular, is the subject of the sentence and has what is called the nominative case. In the sentence 'I threw away the flower', I am now the subject and the flower has become the direct object in the accusative case. In the sentence 'I did not like the colour of the flower', I am again the subject, the colour is now the object and the flower has become a possessive noun and has the genitive case. In the sentence 'The flower fell to the ground' the flower is once again the subject (nominative) and the ground has the dative case. If we add 'with a whisper', then whisper takes the ablative case. In other words, case confers on nouns an expression of their meaning in any sentence. This is shown by the ending of the Latin word, which changes with case and number and, in so doing changes the naked word into part of a sentence (Table 2).

Table 2

| Case | Singular | | Plural | |
|---|---|---|---|---|
| nominative | *flos* | the flower (subject) | *flores* | the flowers |
| accusative | *florem* | the flower (object) | *flores* | the flowers |
| genitive | *floris* | of the flower | *florum* | of the flowers |
| dative | *flori* | to or for the flower | *floribus* | to or for the flowers |
| ablative | *flore* | by, with or from the flower | *floribus* | by, with or from the flowers |

Table 3

| Declension | I | II | | III | | IV | | V |
|---|---|---|---|---|---|---|---|---|
| Gender | f | m | n | m.f | n | m | n | f |
| **Singular** | | | | | | | | |
| nom | -a | -us(er) | -um | * | * | -us | -u | -es |
| acc | -am | -um | -um | -em(im) | * | -um | -u | -em |
| gen | -ae | -i | -i | -is | -is | -us | -us | -ei |
| dat | -ae | -o | -o | -i | -i | -ui(u) | -ui(u) | -ei |
| abl | -a | -o | -o | -e(i) | -i(e) | -u | -u | -e |
| **Plural** | | | | | | | | |
| nom | -ae | -i | -a | -es | -ia | -us | -ua | -es |
| acc | -as | -os | -a | -es(is) | -ia | -us | -ua | -es |
| gen | -arum | -orum | -orum | -ium | -ium | -uum | -uum | -erum |
| dat | -is | -is | -is | -ibus | -ibus | -ibus | -ibus | -ibus |
| abl | -is | -is | -is | -ibus | -ibus | -ibus | -ibus | -ebus |

* Denotes various irregular endings.

Nouns fall into five groups, or declensions, as determined by their endings (Table 3).

Generic names are singular and are treated as subjects, taking the nominative case. *Solanum* means 'Comforter' and derives from the use of nightshades as herbal sedatives. The gender of generic names is that of the original Greek or Latin noun or, if that was variable, is chosen by the author of the name. There are exceptions to this in which masculine names are treated as feminine, and fewer in which compound names, which ought to be feminine, are treated as masculine. As a general guide, names ending in *-us* are masculine unless they are trees (such as *Fagus*, *Pinus*, *Quercus*, *Sorbus*) which are treated as feminine; names ending in *-a* are feminine and names ending in *-um* are neuter; names ending in *-on* are masculine unless they can also take *-um*, when they are neuter, or the ending is *-dendron* when they are also neuter (*Rhododendron* or *Rhododendrum*); names ending in *-ma* (as in terminations such as *-osma*) are neuter; names ending in *-is* are mostly feminine or masculine treated as feminine (*Orchis*) and those ending in *-e* are neuter; other feminine endings are *-ago*, *-odes*, *-oides*, *-ix* and *-es*.

A recommendation for forming generic names to commemorate men or women is that these should be treated as feminine and formed as follows:

> For names ending in a vowel, terminate with *-a*
>> except those ending in *-a*, which terminate with *-ea*
>> and those ending in *-ea*, which should not change.
>
> For names ending in a consonant, add *-ia*
>> except those ending in *-er*, to which add *-a*.
>
> For latinized names ending in *-us*, change the ending to *-ia*.

Generic names which are formed arbitrarily or are derived from vernacular names have their ending selected by the name's author.

Table 4

| Masculine | Feminine | Neuter | |
|---|---|---|---|
| -us | -a | -um | *hirsutus* (hairy) |
| -is | -is | -e | *brevis* (short) |
| -os | -os | -on | *acaulos* ἄκαυλος (stemless) |
| -er | -era | -erum | *asper* (rough) |
| -er | -ra | -rum | *scaber* (rough) |
| -ax | -ax | -ax | *fallax* (false) |
| -ex | -ex | -ex | *duplex* (double) |
| -ox | -ox | -ox | *ferox* (very prickly) |
| -ans | -ans | -ans | *reptans* (creeping) |
| -ens | -ens | -ens | *repens* (creeping) |
| -or | -or | -or | *tricolor* (three-coloured) |
| -oides | -oides | -oides | *bryoides* βρύον, εἶδος (moss-like) |

### Species names

The name of a species is a binary combination of the generic name followed by a specific epithet. If the epithet is of two words they must be joined by a hyphen or united into one word. The epithet can be taken from any source whatever and may also be constructed in an arbitrary manner. It would be reasonable to expect that the epithet should have a descriptive purpose, and there are many which do, but large numbers either refer to the native area in which the plant grows or commemorate a person (often the discoverer, the introducer into cultivation or a noble personage). The epithet may be adjectival (or descriptive), qualified in various ways with prefixes and suffixes, or a noun.

It will become clear that because descriptive adjectival epithets must agree with the generic name, the endings must change in gender, case and number. For example, *Dipsacus fullonum* L. has the generic name used by Dioscorides meaning 'Dropsy', alluding to the accumulation of water in the leaf-bases, and an epithet which is the masculine genitive plural of *fullo*, a fuller, and which

Table 5

| Masculine | Feminine | Neuter | | |
|-----------|----------|--------|---|---|
| *-us* | *-a* | *-um* | *longus* | (long) |
| *-ior* | *-ior* | *-ius* | | (longer) |
| *-issimus* | *-issima* | *-issimum* | | (longest) |
| *-is* | *-is* | *-e* | *gracilis* | (slender) |
| *-ior* | *-ior* | *-ius* | | (slenderer) |
| *-limus* | *-lima* | *-limum* | | (slenderest) |
| *-er* | *-era* | *-erum* | *tener* | (thin) |
| *-erior* | *-erior* | *-erius* | | (thinner) |
| *-errimus* | *-errima* | *-errimum* | | (thinnest) |

identifies the typical form of this teasel as the one which was used to clean and comb up a 'nap' on cloth. The majority of adjectival epithet endings are as in the first two examples listed in Table 4.

Comparative epithets are informative because they provide us with an indication of how the species contrasts with the general features of other members of the genus (Table 5).

### Epithets commemorating people

Specific epithets which are nouns are grammatically independent of the generic name: *Campanula trachelium* is literally 'Little bell' (feminine) 'neck' (neuter). When they are derived from the names of people, they can either be retained as nouns in the genitive case (*clusii* is the genitive singular of Clusius, the latinized version of l'Ecluse, and gives an epithet with the meaning 'of l'Ecluse') or be treated as adjectives and then agreeing in gender with the generic noun (*Sorbus leyana* Willmott is a tree taking, like many others, the feminine gender despite the masculine ending, and so the epithet which commemorates Augustin Ley also takes the feminine ending). The epithets are formed follows:

to names ending with a vowel (except *-a*) or *-er*, add *-i* when masculine singular, *-ae* when feminine singular, *-orum* when masculine plural, or *-arum* when feminine plural

to names ending with *-a*, add *-e* when singular or *-rum* when plural

to names ending with a consonant (except *-er*), add *-ii* when masculine singular, *-iae* when feminine singular, *-iorum* when masculine plural, or *-iarum* when feminine plural

or, when used adjectivally:

to names ending with a vowel (except *-a*), add *-anus* when masculine, *-ana* when feminine, or *-anum* when neuter

to names ending with *-a*, add *-nus* when masculine, *-na* when feminine, or *-num* when neuter

to names ending with a consonant, add *-ianus* when masculine, *-iana* when feminine, or *-ianum* when neuter

## *Geographical epithets*

When an epithet is derived from the name of a place, usually to indicate the plant's native area but also, sometimes, to indicate the area or place from which the plant was first known or in which it was produced horticulturally, it is preferably adjectival and takes one of the following endings:

*-ensis* (m), *-ensis* (f), *-ense* (n)

*-(a)nus* (m), *-(a)na* (f), *-(a)num* (n)

*-inus* (m), *-ina* (f), *-inum* (n)

*-icus* (m), *-ica* (f), *-icum* (n)

Geographical epithets are sometimes inaccurate because the author of the name was in error as to the true origin of the plant, or obscure because the ancient classical names are no longer familiar to us. As with epithets which are derived from proper names to commemorate people, or from generic names or vernacular names which are treated as being Latin, it is now customary to start them with a small initial letter but it remains permissible to give them a capital initial.

*Categories below the rank of species*

Botanically, the subdivision of a species group is based upon a concept of infraspecific variation which assumes that evolutionary changes in nature are progressive fragmentations of the parent species. Put in another way, a species, or any taxon of lower rank, is a closed grouping whose limits embrace all their lower ranked variants (subordinate taxa). It will be seen later that a different concept underlies the naming of cultivated plants, which does not make such an assumption but recognizes the possibility that cultivars may straddle species or other boundaries, or overlap each other, or be totally contained, one by another.

The rules by which botanical infraspecific taxa are named specify that the name shall consist of the name of the parent species followed by a term which denotes the rank of the subdivision, and an epithet which is formed in the same ways as specific epithets, including grammatical agreement when adjectival. Such names are subject to the rules of priority and typification. The ranks form a hierarchy as follows: *subspecies* (abbreviated to *subsp.* or *ssp.*), *varietas* (variety in English, abbreviated to *var.*), *subvarietas* (subvariety or *subvar.*), *forma* (form or *f.*). Further subdivisions are permitted but the Botanical Code does not define the characteristics of any rank within the hierarchy. Consequently, infraspecific classification is subjective.

When a subdivision of a species is named which does not include the nomenclatural type of the species, it automatically establishes the name of the equivalent subdivision which *does* contain that type. Such a name is an 'autonym' and has the same epithet as the species itself but is not attributed to an author. This is the only event which permits the repetition of the specific epithet and the only permissible way of indicating that the taxon includes the type for the species name. The same constraints apply to subdivisions of lower ranks. For example: *Veronica hybrida* L. was deemed by E. F. Warburg to be a component of *Veronica spicata* L. and he named it *V. spicata* L. subsp. *hybrida* (L.) E. F. Warburg. This implies the existence of a typical subspecies, the autonym for which is *V. spicata* L. subsp. *spicata*. It will be seen from the citation of Warburg's new combination that the disappearance of a former Linnaean species can be explained. Retention of the epithet *hybrida*, and the indication of Linnaeus being its author (in brackets) shows the benefit of this system in constructing names with historic meanings.

## Hybrids

Hybrids are particularly important as cultivated plants but are also a feature of many plant groups in the wild, especially woody perennials such as willows. The rules for the names and naming of hybrids are contained in the Botanical Code but are equally applicable to cultivated plant hybrids.

For the name of a hybrid between parents from two different genera, a name can be constructed from the two generic names, in part or in entirety (but not both in their entirety) as a condensed formula. For example, × *Mahoberberis* is the name for hybrids between the genera *Mahonia* and *Berberis* (in this case the cross is only bigeneric when *Mahonia*, a name conserved against *Berberis*, is treated as a distinct genus), and × *Fatshedera* is the name for hybrids between the genera *Fatsia* and *Hedera*. The orchid hybrid between *Gastrochilus bellinus* (Rchb. f.) O. Ktze. and *Doritis pulcherrima* Lindl. carries the hybrid genus name × *Gastritis* (it has a cultivar called 'Rumbling Tum'!). Alternatively, a formula can be used in which the names of the genera are linked by the sign for hybridity ' × ': *Mahonia* × *Berberis* and *Fatsia* × *Hedera*. Hybrids between parents from three genera are also named either by a formula or a condensed formula, and in all cases the

41

condensed formula is treated as a generic name if it is published with a statement of parentage. When published, it becomes the correct generic name for any hybrids between species of the named parental genera. A third alternative is to construct a commemorative name in honour of a notable person and to end it with the termination *-ara*: × *Sanderara* is the name applied to the orchid hybrids between the genera *Brassia*, *Cochlioda* and *Odontoglossum* and commemorates H. F. C. Sander, the British orchidologist.

A name formulated to define a hybrid between two particular species from different genera can take the form of a species name, and then applies to all hybrids produced subsequently from those parent species: × *Fatshedera lizei* Guillaumin is the name first given to the hybrid between *Fatsia japonica* (Thunb.) Decne. & Planch. and *Hedera helix* L. cv. Hibernica, but which must include all hybrids between *F. japonica* and *H. helix* and × *Cupressocyparis leylandii* (Jackson & Dallimore) Dallimore is the name for hybrids between *Chamaecyparis nootkatensis* (D. Don) Spach and *Cupressus macrocarpa* Hartweg ex Godron. Because the parents themselves are variable, the progeny of repeated crosses may be distinctive and so warrant naming. They may be named under the Botanical Code (prior to 1983 they would have been referred to as nothomorphs or bastard forms) and also under the International Code of Nomenclature for Cultivated Plants as 'cultivars': for example, × *Cupressocyparis leylandii* cv. Naylor's Blue. The hybrid nature of × *Sanderara* is expressed by classifying it as a 'nothogenus' (bastard genus, or, in the special circumstances of orchid nomenclature, grex class) and of × *Cupressocyparis leylandii* by classifying it as a 'nothospecies' (within a nothogenus). For infraspecific ranks the multiplication sign is not used but the term denoting their rank receives the prefix notho-, or 'n-' (*Mentha* × *piperita* L. nothosubspecies *pyramidalis* (Ten.) Harley which, as stated earlier, also implies the autonymous *Mentha* × *piperita* nothosubspecies *piperita*).

Hybrids between species in the same genus are also named by a formula or by a new distinctive epithet: *Digitalis lutea* L. × *D. purpurea* L. and *Nepeta* × *faassenii* Bergmans ex Stearn are both correct designations for hybrids. In the example of *Digitalis*, the order in which the parents are presented happens to be the correct order, with the seed parent first. It is permissible to indicate the roles of the parents by including the symbols for female '♀' and male '♂', when this information is known, or otherwise to present the parents in alphabetical order.

The orchid family presents particularly complex problems of nomenclature, requiring its own 'Code' in the form of the *Handbook on Orchid Nomenclature and Registration* (International Orchid Commission, 1985). There are some 20000 species of orchids and to these have been added a huge range of hybrids (some with eight genera contributing to their parentage) and over 70000 hybrid swarms, or grexes, with a highly complex ancestral history.

Sometimes a hybrid is sterile because the two sets of chromosomes which it has inherited, one from each parent, are sufficiently dissimilar to cause breakdown of the mechanism which ends in the production of gametes. In such cases doubling its chromosome complement may produce a new state of sexual fertility and what is, in effect, a new biological species. Many naturally occurring species are thought to have evolved by such changes and man has created others artificially via the same route, some intentionally and some unintentionally from the wild. The bread-wheats, *Triticum aestivum* L. are an example of the latter. They are not known in the wild and provide an example of a complex hybrid ancestry but whose name does not need to be designated as hybrid. Even an artificially created tetraploid (having, as above, four instead of the normal two sets of chromosomes) need not be designated as hybrid by inclusion of ' × ' in its name: *Digitalis mertonensis* Buxton & Darlington is the tetraploid from an infertile hybrid between *D. grandiflora* L. and *D. purpurea* L.

## *Synonymy and illegitimacy*

Inevitably, most plants have been known by two or more names in the past. Since a plant can have only one correct name, which is determined by priority, its other validly published names are synonyms. A synonym may be one which is strictly referable to the same type (a nomenclatural synonym) or one which is referable to another type which is, however, considered to be part of the same taxon (this is a taxonomic synonym). The synonymy for any plant or group of plants is important because it provides a reference list to the history of the classification and descriptive literature on that plant or group of plants.

In the search for the correct name, by priority, there may be names which have to be excluded from consideration because they are regarded as being illegitimate, or not in accordance with the rules.

Names which have the same spelling but are based on different types from that which has priority are illegitimate 'junior homonyms'. Clearly, this prevents the same name being used for different plants. Curiously, this exclusion also applies to the names of those animals which were once regarded as plants, but not to any other animal names.

Names published for taxa which are found to include the type of an existing name are illegitimate because they are 'superfluous'. This prevents unnecessary and unacceptable proliferation of names of no real value.

Names of species in which the epithet exactly repeats the generic name have to be rejected as illegitimate 'tautonyms'. It is interesting to note that there are many plant names which have achieved some pleonastic repetition by using generic names with Greek derivation and epithets with Latin

44

derivation – for example, *Arctostaphylos uva-ursi* (bear-berry, berry of the bear), *Myristica fragrans* (smelling of myrrh, fragrant), *Orobanche rapum-genistae* (legume strangler, rape of broom); or the reverse of this, *Liquidambar styraciflua* (liquid amber, flowing with storax) – but modern practice is to avoid such constructions. In the animal kingdom tautonyms are commonplace.

The Code provides a way of reducing unwelcome disturbance to customary usage which would be caused by rigid application of the rule of priority to replace any incorrect names. Certain names of families and genera which, although incorrect or problematic are, for various reasons (usually their long usage and wide currency in important literature) may be agreed to be conserved at a Botanical Congress. These conserved names can be found listed in an Appendix to the Code, together with names which are to be rejected because they are taxonomic synonyms used in a sense which does not include the type of the name, or are earlier nomenclatural synonyms based on the same type, or are homonyms or orthographic variants.

The Code also recommends the ways in which names should be spelt or transliterated into Latin form in order to avoid what it refers to as 'orthographic variants'. The variety found amongst botanical names includes differences in spelling which are, however, correct because their authors chose the spellings when they published them and differences which are not correct because they contain any of a range of defects which have become specified in the Code. This is a problem area in horticultural literature, where such variants are commonplace. It is clearly desirable that a plant name should have a single, constant and correct spelling but this has not been achieved in all fields and reaches its worst condition in the labelling of plants for sale in some nurseries.

# The International Code of Nomenclature for Cultivated Plants

In 1952, the Committee for the Nomenclature of Cultivated Plants of the International Botanical Congress and the International Horticultural Congress in London adopted the International Code of Nomenclature for Cultivated Plants. Sometimes known as the Cultivated Code, it was first published in 1953 and has been revised at intervals since then (Brickell *et al.*, 1980). This Code formally introduced the term 'cultivar' to encompass all varieties or derivatives of wild plants which are raised under cultivation. The aim of the Code is to 'promote uniformity and fixity in the naming of agricultural, sylvicultural and horticultural cultivars (varieties)'. There can be no doubt that the diverse approaches to naming garden plants, by common names, by botanical names, by mixtures of botanical and common names, by group names and by fancy names, is no less complex than the former unregulated use of common or vernacular names.

Despite the various ways of naming cultivated plants, the psychology of advertising takes descriptive naming into new dimensions. It catches the eye with bargain offers of colourful, vigorous and hardy, large-headed, incurved *Chrysanthemum* cvs. by referring to them as HARDY FOOTBALL MUMS. However, we are not here concerned with the ethics of mail-order selling techniques but with the regulation of meaningful names under the Code.

This Code governs the names of all plants which retain their distinctive characters when reproduced sexually (by seed) or vegetatively in cultivation. Because the Code does not have legal status, the commercial interests of plant breeders are guarded by the Council of the International Union for the Protection of New Varieties of Plants (UPOV). In Britain, the Plant Varieties Rights Office works with the Government to have UPOV's guidelines implemented.

The Cultivated Code accepts the rules of botanical nomenclature and the retention of the botanical names of those plants which are taken into cultivation from the wild. It recognizes only the one category of garden-maintained variant, the cultivar (cv.) or garden variety, which should not be confused with the botanical *varietas*. It recognizes also the supplementary, collective category of the 'group', intermediate between species and cultivar, for special circumstances explained below. Unlike wild plants, cultivated plants receive unnatural treatment and selection pressures from man and are maintained by him. The term cultivar covers:

    clones which are derived vegetatively from a single parent,

    lines of selfed or inbred individuals,

    series of cross-bred individuals, and

    assemblages of individuals which are resynthesized only by cross-breeding (e.g. $F_1$ hybrids).

From this it will be seen that because of the single category of cultivar, the hybrid between parents of species rank or any other rank has equal status with a 'line' selected within a species or taxon of any other rank, including another cultivar, and that parity exists only between names, not between biological entities. The creation of a cultivar name does not, therefore, reflect a fragmentation of the parent taxon but does reflect the existence of a group of plants having a particular set of features, without definitive reference to its parents. Features may be concerned with cropping, disease resistance or biochemistry, showing that the Cultivated Code requires a greater flexibility than the Botanical Code. It achieves this by having no limiting requirement for 'typical' cultivars but by regarding cultivars as part of an open system of nomenclature. Clearly, this permits a wide range of applications and differences from the Botanical Code and these are considered in a recent publication (Styles, 1986).

Since January 1st, 1959, the names of cultivars have had to be 'fancy names' in common language and not in Latin. Fancy names come from any source. They can commemorate anyone, not only persons connected with botany or plants, or they can identify the nursery of their origin, or be descriptive, or be truly fanciful. Those which had Latin garden variety names were allowed to remain in use: *Nigella damascena* L. has two old varietal names *alba* and *flore pleno* and also has a modern cultivar with the fancy name cv. Miss Jekyll. In the glossary, no attempt has been made to include fancy names but a few of the earlier Latin ones have been included.

In order to be distinguishable, the fancy names have to be printed in a type-face unlike that of the species name and to be given capital initials. They also have to be either preceded by 'cv.', as above, or placed between single quotation marks. Thus, *Salix caprea* L. cv. Kilmarnock, or 'Kilmarnock' is a weeping variety of the goat willow and is also part of the older variety *Salix caprea* var. *pendula*; other examples are *Geranium ibericum* Cav. cv. Album and *Acer davidii* Franchet 'George Forrest'.

Fancy names can be attached to an unambiguous common name, such as potato 'Duke of York' for *Solanum tuberosum* L. cv. Duke of York or to a generic name such as *Cucurbita* 'Table Queen' for *Cucurbita pepo* L. cv. Table Queen, or of course to the botanical name, even when this is below the rank of species, *Rosa sericea* var. *omeiensis* 'Praecox'. However, the same fancy name may not be used twice within a group (cultivar class) if such duplication would cause ambiguity. Thus, cherries and plums are in distinct cultivar classes and we would never refer to either by the generic name, *Prunus*, alone. Consequently, the same fancy name could be used for a cultivar of a cherry and for a cultivar of a plum: cherry 'Early Rivers' and plum 'Early Rivers' (now called 'Rivers Early Prolific').

For some extensively bred crops and decorative plants there is a long-standing supplementary category: the group. By naming the group in such plants, a greater degree of accuracy is given to the garden name: for example, pea (wrinkle-seeded group) 'Laxton's Progress' and *Rosa* (rambler) 'Alberic Barbier' and *Rosa* (rugosa) 'Agnes'.

To ensure that a cultivar has only one correct name, the Cultivated Code requires that priority acts and, to achieve this, publication and registration are necessary. As with botanical names, cultivars can have synonyms and the problems are increased in this respect because it is permissible to translate the fancy names into other languages. To establish a fancy name, publication has to be in printed matter which is dated and distributed to the public. For the more popular groups of plants, usually genera, there are societies which maintain statutory registers of names, and the plant breeding industry has available to it the Plant Variety Rights Office as a statutory body for registration of crop-plant names as trade marks for commercial protection, including patent rights on vegetatively propagated cultivars.

## Chimaeras

One group of plants which is almost entirely within the province of gardening is that of the graft chimaeras, or graft hybrids. These are plants in which a mosaic of tissues from the two parents in a grafting partnership results in an individual plant, upon which shoots resembling each of the parents, and in some cases shoots of intermediate character, are produced in an unpredictable manner. Unlike sexually produced hybrids, the admixture of the two parents' contributions is not at the level of the nucleus in each and every cell but is more like a marbling of a ground tissue of one parent with streaks of tissue of the other parent. Chimaeras can also result from mutation in a growing point, from which organs are formed composed of normal and mutant tissues, as with genetic forms of variegation. In all cases, three categories may be recognized, called sectorial, mericlinal and periclinal chimaeras, in terms of the extent of tissue 'marbling'. The chimaeral condition is denoted by the addition sign ' + ', instead of the multiplication sign ' × ' used for true hybrids. A chimaera which is still fairly common in Britain is that named +*Laburnocytisus adamii* C. K. Schneider. This was the result of a graft between *Cytisus purpureus* Scop. and *Cytisus laburnum* L., which are now known as *Chamaecytisus purpureus* (Scop.) Link and *Laburnum anagyroides* Medicus, respectively. Although its former name *Cytisus + adamii* would not now be correct, the name *Laburnocytisus* meets the requirement of combining substantial parts of the two parental generic names, and can stand.

It is interesting to speculate that if cell-culture and callus-culture techniques could be used to produce chimaeral mixtures to order, it may be possible to create some of the conditions which were to have brought about the 'green revolution'. Protoplast fusion methods failed to combine the culturally and economically desirable features of distant parents, which were to have given multi-crop plants and new nitrogen-fixing plants, because of the irregularities in fusion of both protoplasts and their nuclei. It may be that intact cells would prove easier to admix. However, a new science of molecular genetics is in the ascendant; this has shown itself capable of modifying the genetic control system in a way which suggests that any aspect of a plant can, potentially, be manipulated to suit man's requirements. The products of genetic manipulation may themselves require nomenclature consideration in the future.

# Botanical terminology

There is nothing accidental about the fact that in our everyday lives we communicate at two distinct levels. Our 'ordinary' conversation employs a rich, dynamic language in which meaning can differ from one locality to another and change from time to time. Our 'ordinary' reading is of a written language of enormous diversity – ranging from contemporary magazines which are intentionally erosive of good standards, to high-quality prose of serious writers. However, when communication relates to specific topics, in which ambiguity is an anathema, the language which we adopt is one in which 'terminology' is relied upon to convey information accurately and incontroversially. Thus, legal, medical and all scientific communications employ terms which have widely accepted meanings and which therefore convey those meanings in the most direct way. Because these terms are derived predominantly from classical roots and have long-standing acceptance, as are botanical terms for the parts of the flower, they have the added advantage of international currency.

This glossary contains many examples of words which are part of botanical terminology as well as being employed as descriptive elements of plant names. Such is the wealth of this terminology that it would make tedious reading to attempt here to discriminate between and explain all the terms relating, say, to the surface of plant leaves and the structures (hairs, glands and deposits) which subscribe to that texture. However, terms which refer to such con-spicuous attributes as leaf shape and the form of inflorescences are very commonly used in plant names and, since unambiguous definition would be lengthy, are illustrated as figures.

More extensive glossaries of terminology can be found in textbooks and floras, but the sixth edition (1955) of Willis' *Dictionary of Flowering Plants and Ferns* (1931) is a particularly rewarding source of information.

## *The glossary*

The glossary is for use in finding out the meanings of the names of plants. There are many plant names which cannot be interpreted or which yield very uninformative translations. Authors have not always used specific epithets with a single, narrow meaning so that there is a degree of latitude in the translation of many epithets. Equally, the spelling of epithets has not remained constant (for example in the case of geographic names), and their variants from one species to another are all correct if they were published in accordance with the Code. In certain groups such as garden plants from, say, China and exotics such as many members of the profuse orchid family, commemorative names have been applied to plants more frequently than in most other groups. If you wish to add further significance to such names, you will find it mostly in literature on plant-hunting and hybridization or from reference works such as that on taxonomic literature by Stafleu and Cowan (1976–).

Generic names in the European flora are mostly of ancient origin. Their meanings, even of those which are not taken from mythological sources, are seldom clear and many have had their applications changed and are now used as specific epithets. Generic names of plants discovered throughout the world in recent times have mostly been constructed to be descriptive and will yield to translation. The glossary contains the generic names of a wide range of both garden and wild plants and treats them as singular nouns, with capital initials. Orthographic variants have not been sought out but a few are presented and have the version which is generally incorrect in brackets. Listings of generic names can be found in Farr (1979–86).

As an example of how the glossary can be used, we can consider the name *Sarcococca ruscifolia*. This is the name given by Stapf to plants which belong to Lindley's genus *Sarcococca* of the family Buxaceae, the box family. In the glossary we find *sarc-*, *sarco-* meaning fleshy and *-coccus -a -um* meaning berried and from this we conclude that *Sarcococca* means Fleshy-berry (the generic name being a singular noun) and has the feminine gender. We also find *rusci-* meaning butcher's-broom-like or resembling *Ruscus* and *-folius -a -um* meaning -leaved and we conclude that this species of Fleshy-berry has leaves which resemble the prickly cladodes (leaf-like branches) of *Ruscus*. The significance of this generic name lies in the fact that dry fruits are more typical in members of the box family than fleshy ones.

From this example, it should be clear that names can be constructed from adjectives or adjectival nouns to which prefixes or suffixes can be added, thus giving them further qualification. As a general rule, epithets which are formed in this way have an acceptable interpretation when '-ed' is added to the English translation: this would render *ruscifolia* as *Ruscus*-leaved.

It will be noted that *Sarcococca* has a feminine ending *-a* and that *ruscifolia* takes the same gender. However, if the generic name had been of the masculine gender the epithet would have become *ruscifolius* and if of the neuter gender then it would have become *ruscifolium*. For this reason the entries in the glossary are given all three endings which, as pointed out earlier, mostly take the form *-us -a -um* or *-is -is -e*.

Where there is the possibility that a prefix which is listed could lead to the incorrect translation of some epithet, the epithet in question is listed close to the prefix and to an

example of an epithet in which the prefix is employed.
Examples are:

*aer-* meaning air- or mist-, gives *aerius -a -um* meaning airy or lofty.

*aeratus -a -um*, however, means bronzed (classically, made of bronze).

*caeno-*, from the Greek *cainos*, means fresh-, but

*caenosus -a -um*, is from the Latin *caenum* and means muddy or growing on mud or filth.

Examples will be found of words which have several fairly disparate meanings. A few happen to reflect differences in meaning of closely similar Greek and Latin source words as in the example above, and others reflect what is to be found in literature, in which other authors have suggested meanings of their own. Similarly, variations in spelling are given for some names and these are also to be found in the literature, although not all of them are strictly permissible for nomenclatural purposes. Their inclusion emphasizes the need for uniformity in the ways in which names are constructed and provides a small warning that there are in print many deviant names, some intentional and some accidental.

Many of the epithets which may cause confusion are either classical geographic names or terms which retain a more specific meaning of the classical language. There are many more such epithets than are listed in this glossary.

# The glossary

*a-, ab-*   away from-, downwards-, without-, un-, very-
*ac-, ad-, af-, ag-, al-, an-, ap-, ar-, as-, at-*   near-, towards-
*abbreviatus -a -um*   shortened
*Abelia*   for Dr Clarke Abel, writer on China (1780–1826)
*Abies*   Rising one (tall tree)
*-abilis -is -e*   -able, -capable of (preceded by some action)
*abnormis -is -e*   departing from normal in some structure
*abortivus -a -um*   with missing or malformed parts
*abros*   delicate
*abrotanoides*   *Artemisia*-like (from an ancient Greek name for wormwood or mugwort)
*abruptus -a -um*   ending suddenly, blunt-ended
*abscissus -a -um*   cut off
*absinthius -a -um*   from an ancient Greek or Syrian name for wormwood
*abyssinicus -a -um*   of Abyssinia, Abyssinian
*Acacia*   Thorn (from the Greek *akis*)
*Aceana*   Thorny-one
*acantho-*   thorny-, spiny-
*Acanthus*   Prickly-one
*acaulis -is -e, acaulos -os -on*   lacking an obvious stem
*accicus -a -um*   with a small acute apical cleft, emarginate
*Acer*   Sharp, the Latin name for a maple (either from its use for lances or its leaf), etymologically linked to oak and acre
*acer, acris, acre*   sharp-tasted, acid
*Aceras*   without a spur, not horned
*acerbus -a -um*   harsh-tasted
*aceroides*   maple-like
*acerosus -a -um*   pointed, needle-like
*acetabulosus -a -um*   saucer-shaped
*acetosus -a -um*   acid, sour
*acetosellus -a -um*   slightly acid
*-aceus -a -um*   -resembling (preceded by a plant name)

*Achillea*   after the Greek warrior Achilles

*Achyranthes*   Chaff-flower

*acicularis -is -e*   needle-shaped

*aciculatus -a -um*   finely marked as with needle scratches

*aciculus -a -um*   sharply pointed (e.g. leaf-tips)

*acidosus -a -um*   acid, sharp, sour

*acinaceus -a -um, aciniformis -is -e*   scimitar-shaped

*acinifolius -a -um*   *Acinos*-leaved, basil-thyme-leaved

*Acinos*   Dioscorides' name for a heavily scented calamint

*acmo-*   pointed- (followed by a part of a plant)

*aconiti-*   aconite-

*Aconitum*   the name used by Theophrastus (poisonous Aconite)

*Acorus*   Without-pupil. Dioscorides' name for an iris (its use in treating cataract)

*acro-*   towards the top-, highest- (followed by a noun, e.g. hair, or a verb, e.g. fruiting)

*Acroceras*   Upper-horn (the glumes have an excurrent vein at the tip)

*Actaea*   from the Greek name for elder (the shape of the leaves)

*actinius -a -um*   sea-anemone-like

*actino-*   radiating- (followed by a part of a plant)

*Actinocarpus*   Radiate-fruit (the spreading ripe carpels of thrumwort)

*aculeatus -a -um*   having prickles, prickly, thorny

*aculeolatus -a -um*   having small prickles or thorns

*acuminatus -a -um*   with a long, narrow and pointed tip (see Fig. 7(c), acuminate

*acuminosus -a -um*   with a conspicuous long flat pointed apex

*acutus -a -um, acuti-*   acutely pointed, sharply angled at the top

*adamantinus -a -um*   from Diamond Lake, Oregon, USA

*Adansonia*   for Michel Adanson, French botanist (1727–1806)

*aden-, adeno-*   gland-, glandular-

*adenotrichus -a -um*   glandular-hairy

*Adiantum*   Unwetted (the old Greek name refers to its staying unwetted under water)

*Adlumia*   for Maj. John Adlum, American viticulturist (1759–1836)

*admirabilis -is -e*   to be admired, admirable

*adnatus -a -um*   joined together, adnate

*Adonis*   see Anemone

*Adoxa*   Gloryless, Without-glory

*adpressus -a -um*   pressed together, lying flat against (e.g. the hairs on the stem) *adscendens*   curving up from a prostrate base, half-erect

*adstringens*   constricted

*adsurgens*   rising up, ascending

*adulterinus -a -um*   of adultery (intermediate between two other species suggesting hybridity, as in *Asplenium adulterinum*)

*aduncus -a -um*   hooked, having hooks

*Aegopodium*   Goat's foot (the leaf shape)

*aemulus -a -um*   imitating, rivalling

*aeneus -a -um*   bronzed

*aequalis -is -e, aequali-*   equal, equally-

*aequilateralis -is -e, aequilaterus -a -um*   equal-sided

*aequinoctialis -is -e*   of the equinox (the flowering time)

*aer-*   air-, mist-

*aeranthos -os -on, Aeranthus*   air-flower (rootless epiphyte)

*aeratus -a -um*   bronzed

*aerius -a -um*   airy, lofty

*aeruginosus -a -um*   rusty, verdigris-coloured

*aeschyno-*   shy-, to be ashamed-

*Aesculus*   Linnaeus' name from the Roman name of an edible acorn

*aestivalis -is -e*   of summer

*aestivus -a -um*   developing in the summer

*Aethionema*   Strange-filaments (those of the long filaments are winged and toothed)

*aethiopicus -a -um*   of Africa, African, of north-east Africa

*Aethusa*   Burning one (its pungency)

*aetnensis -is -e*   from Mt Etna, Sicily

*-aeus*   -belonging to (of a place)

*afer, afra, afrum*   African

*affinis -is -e*   related, similar to

*aflatunensis -is -e*   from Aflatun, central Asia

*Afzelia*   for Adam Afzelius, Swedish botanist (1750–1837)

*Agapanthus*   Love-flower

*agastus -a -um*   charming, pleasing

*Agathelpis*   Good-hope (its natural area on the Cape)

*Agathis*   Ball-of-twine (the appearance of the cones)

*Agave*   Admired-one

*Ageratum*   Unageing (the flower-heads retain their colour for a long period)

*agetus -a -um*   wonderful

*agglomeratus -a -um*   in a close head, congregated together

*agglutinatus -a -um*   glued or firmly joined together

*aggregatus -a -um*   clustered together

*-ago*   -like

*agrarius -a -um, agrestis -is -e*   of fields, wild on arable land

*Agrimonia*   Cataract (from its medicinal use)

*Agropyron(um)*   Field-wheat

*Agrostemma*   Field-garland (Linnaeus' view of its suitability for such)

*Agrostis*   Field (-grass)

*Ailanthus*   Reaching-to-heaven (from a Moluccan name)

*Aira*   old Greek name for darnel grass

*Aitonia*   for Wm. Aiton, head gardener at Kew (1759–1793)

*Aizoon*   Always-alive

*ajacis -is -e*   of Ajax (from whose blood grew a flower marked with AIA)

*ajanensis -is -e*   from Ajan, East Asia

*Ajuga*   corrupted Latin for Abortifant

*alabastrinus -a -um*   like alabaster

*alaternus*   an old generic name for a buckthorn

*alatus -a -um, alati-, alato-*   winged, with protruding ridges which are wider than thick

*albatus -a -um*   turning white

*albens*   white

*Alberta*   for Albertus Magnus (1193–1280) (*A. magna* is from Natal)

*albertii, albertianus -a -um*   for Albert, Prince Consort, or for Dr Albert Regel, Russian plant collector in Turkestan

*albescens*   turning white

*albicans, albidus -a -um, albido-, albulus -a -um*   whitish

*Albizia (Albizzia)*   for Filippo degli Albizzi, Italian naturalist

*Albuca*   White

*albus -a -um, albi-, albo-*   dead-white

*alceus -a -um*   mallow-like, from 'alcea', the name used by Dioscorides

*Alchemilla*   from Arabic reference either to reputed magical properties or to fringed leaves of some species

*alcicornis -is -e*   elk-horned

*aleppicus -a -um*   of Aleppo, North Syria

*aleur-, aleuro-*   mealy-, flowery- (surface texture)

*aleuticus -a -um*   Aleutian

*algidus -a -um*   cold, of high mountains

*alicae*   for Princess Alice of Hess (1843–1878)

*alicia*   for Miss Alice Pegler, plant collector in Transkei, South Africa

*-alis -is -e*   -belonging to (a place)

*Alisma*   Dioscorides' name for a plantain-leaved water plant

*alkakengi*   a name used by Dioscorides

*Allamanda*   for Dr Allamand, who sent seeds of this to Linnaeus, from Brazil

*allantoides*   sausage-shaped

*allatus -a -um*   introduced, not native

*alliaceus -a -um, alloides*   *Allium*-like, smelling of garlic

*Alliaria*   Garlic-smelling

*allionii*   for Carlo Allioni, author of *Flora Pedemontana* (1705–1804)

*Allium*   the ancient Latin name for garlic

*allo-*   diverse-, several-, different-, other-

*Allosorus*   Variable-sori (their shapes vary)

*almus -a -um*   bountiful

*alni-*   *Alnus*-like-, alder-like-

*Alnus*   the Latin name for the alder

*Alonsoa*   for Alonzo Zanoni, Spanish official in Bogotá

*Alopecurus*   Fox's tail

*alpester -ris -re*   of mountains

*alpicolus -a -um*   of high mountains

*Alpinia*   for Prosper Alpino, Italian botanist (1553–1616)

*alpinus -a -um*   alpine, of mountain pastures

*alsaticus -a -um*   from Alsace, France

*Alsine*   a name used by Dioscorides for a chickweed-like plant

*also-*   leafy-, of groves-

*Alstonia*   for Prof. Charles Alston, of Edinburgh (1716–1760)

*alstonii*   for Capt. E. Alston, collector of succulents in Ceres, South Africa

*alternans*   alternating

*Althaea*   Healer, a name used by Theophrastus

*alti, alto-, altus -a -um*   tall, high

*altilis -is -e*   nutritious, fat, large

*alumnus -a -um*   well-nourished, flourishing

*alutaceus -a -um*   of the texture of soft leather

*alveolatus -a -um*   with shallow pits, alveolar

*Alyssum*   Pacifier, Without fury, Not rage

*amabilis -is -e*   pleasing, lovely

*Amaranthus (Amarantus)*   Unfading

*amaranticolor*   purple, *Amaranthus*-coloured

*amarellus -a -um, amarus -a -um*   bitter (as in the amaras or bitters of the drinks industry, e.g. *Quassia amara*)

*Amaryllis*   the name of a country girl in Latin writings

*amaurus -a -um*   dark, without lustre

*amazonicus -a -um*   from the Amazon basin, South America

*amb-, ambi-*   around-

*ambigens, ambiguus -a -um*   doubtful, of uncertain relationship

*ambly-*   blunt-

*amblyodon*   blunt-toothed

*amboinensis -is -e*   from Amboina, Indonesia

*ambrosia*   elixir of life, food of the gods

*Amelanchier*   a Provençal name for snowy-mespilus

*amentaceus -a -um*   having catkins, catkin bearing

*amesianus -a -um*   for Frederick Lothrop Ames, American orchidologist, or for Prof Oakes Ames of Harvard Botanic Garden and orchidologist

*amethystea*   the colour of amethyst gems

*amicorum*   of the Friendly Isles, Tongan

*amictus -a -um*   clad, clothed

*amiculatus -a -um*   cloaked, mantled

*Ammi*   Sand, a name used by Dioscorides

*Ammophila*   Sand-lover

*ammophillus -a -um*   sand-loving (the habitat)

*amoenus -a -um*   pleasing, delightful

*amomum*   purifying (the Indian spice plant, *Amomum* was used to cure poisoning)

*ampelo-*   wine-, vine-

*Ampelopsis*   Vine-resembler

*amphi, ampho-*   on-both-sides, in-two-ways-, both-, double-

*amphibius -a -um*   with a double life, growing both on land and in water

*amplectans*   stem-clasping (leaves)

*amplexicaulis -is -e*   embracing the stem (e.g. the base of the leaf, see Fig. 6(*d*))

*amplissimus -a -um*   very large, the biggest

*amplus -a -um*   large

*ampullaceus -a -um, ampullaris -is -e*   flask-shaped

*amurensis -is -e*   from the region of the Amur river, eastern Siberia

*amygdalinus -a -um*   of almonds, almond-like

*Amygdalus*   the Greek name for the almond-tree

*an-, ana-*   up-, upon-, upwards-, without-, backwards-, above-, again-

*Anacamptis*   Bent back (the long spur of the flower)

*Anacardium*   Heart-shaped (Linnaeus' name refers to the shape of the false-fruit)

*Anacharis*   Without-charm

*Anagallis*   Unpretentious, Without-boasting, Without-adornment

*anagyroides*   resembling *Anagyris*, curved backwards

*Ananas*   probably of Peruvian origin

*Anaphalis*   Greek name for an immortelle

*anastaticus -a -um*   rising up (*Anastatica hierochuntica*, resurrection plant or rose of Jericho)

*anatolicus -a -um*   from Anatolia, Turkish

*anceps*   two-edged (stems), two-headed

*Anchusa*   Strangler (astringent properties, Greek name formerly for a plant yielding a red dye)

*ancistro-*   fish-hook-

*ancylo-*   hooked-

*andegavensis -is -e*   from Angers in Anjou, France

*Andersonia*   for W. Anderson, botanist on Cook's second and third voyages

*andersonii*   for Thos. Anderson, botanist in Bengal, or for J. Anderson who collected in the Gold Coast

*andicolus -a -um*   from the central Andean cordillera

*andro-, andrus -a -um*   male, stamened, anthered

*androgynus -a -um*   with staminate and pistillate flowers on the same head

*Andromeda*   after the daughter of Cepheus and Cassiope recued by Perseus from the sea monster

*Andropogon*   Bearded-male (awned male spikelet)

*Androsaemum*   Man's blood (the blood-coloured juice of the berries)

*Anemone*   A name used by Theophrastus. Possibly a corruption of Naaman, a Semitic name for Adonis whose blood gave rise to the crimson *Anemone coronaria*. It could also be a corruption of an invocation to the goddess of retribution, Nemesis. Commonly called Windflower.

*anfractuosus -a -um*   twisted, bent

*Angelica*   from the Latin for an angel (healing powers, see *Archangelica*)

*angio-*   urn-, vessel-, enclosed-, (boxed)

*anglicus -a -um, anglicorum*   of the English, English

*anguinus -a -um*   serpentine, snake-like (shape)

*angularis -is -e*   angular

*anguligerus -a -um*   having hooks

*angulosus -a -um*   having angles, angular

*angusti-, angustus -a -um*   narrow

*angustior*   narrower

*Anisantha*   Unequal-flower (flowers vary in their sexuality)

61

*anisatus -a -um*   aniseed-scented

*aniso-*   unequally-, unequal-, uneven-

*Anisophyllea*   Unequal-leaf (the alternate large and small leaves)

*ankylo-*   crooked-

*Annona* (*Anona*)   from the Haitian name

*annotinus -a -um*   one year old, of last year (with distinct annual increments)

*annularis -is -e*   ring-shaped

*annuus -a -um*   annual

*ano-*   upwards-, up-, towards the top-

*Anogramma*   Towards-the-top-lined (sori mature at the tips of the pinnae first)

*anomalus -a -um*   unlike its allies, out of the ordinary

*anopetalus -a -um*   erect-petalled

*anosmus -a -um*   without fragrance, scentless

*ansatus -a -um, ansiferus -a -um*   having a handle

*anserinus -a -um*   of the goose, of the meadows

*ante-*   before-

*Antennaria*   Feeler (the hairs of the pappus)

*anthelminthicus -a -um*   vermifuge, worm expelling

*Anthemis*   Flowery (name used by Dioscorides)

*-anthemus -a -um, -anthes*   -flowered

*Anthericum*   the Greek name for an asphodel

*antherotes*   brilliant

*antho-*   flower-

*anthora*   resembling *Ranunculus thora* in poisonous properties

*Anthoxanthum*   Yellow flower (the mature spikelets)

*anthracinus -a -um*   coal-black

*Anthriscus*   from a Greek name for another umbellifer

*anthropophagorus -a -um*   of the man-eaters

*anthropophorus -a -um*   man-bearing (flowers of the man orchid)

*Anthurium*   Flower tail (the long spadix)

*-anthus -a -um*   -flowered

*Anthyllis*   Downy flower (hair on the calyx)

*anti-*   against-, opposite-, opposite-to-, like-

*anticus -a -um*   turned inwards towards the axis

*antidysentericus -a -um*   against dysentery (use in medical treatment)

*antillarus -a -um*   from the Antilles, West Indies

*antipyreticus -a -um*   against fire (the moss *Fontinalis antipyretica* was packed around chimneys to prevent thatch from igniting

*antiquorum*   of the ancients

*antiquus -a -um*   ancient

*Antirrhinum*   Nose-like
*an(n)ulatus -a -um*   with rings, ringed
*-anus -a -um*   -belonging to, -having
*anvegadensis -is -e*   see *andegavensis*
*ap-, apo-*   without-, away from-, downwards-
*aparine*   a name used by Theophrastus (clinging, seizing)
*apenninus -a -um*   of the Italian Apennines
*Apera*   a meaningless name used by Adanson
*apertus -a -um*   open, bare, naked
*aphaca*   a name used in Pliny for a lentil-like plant
*Aphanes*   Inconspicuous (unnoticed)
*aphyllus -a -um*   without leaves, leafless (perhaps at flowering time)
*apiatus -a -um*   bee-like, spotted
*apicatus -a -um*   with a pointed tip
*apiculatus -a -um*   with a small broad point at the tip, apiculate (see Fig. 7 (e))
*apifer -era -erum*   bee-like, bee-bearing (flowers of the bee orchid)
*Apium*   a name used in Pliny for a celery-like plant. Some relate it to the Celtic 'apon', water, as its preferred habitat
*apodectus -a -um*   acceptable
*apodus -a -um*   without a foot, stalkless
*Aponogeton*   Without neighbour (see *Potamogeton*)
*appendiculatus -a -um*   with appendages
*applanatus -a -um*   flattened out
*appressus -a -um*   lying close together, adpressed
*appropinquatus -a -um*   near, approaching (resemblance to another species)
*apricus -a -um*   sun-loving
*apterus -a -um*   without wings, wingless
*aquaticus -a -um, aquatilis -is -e*   growing in or under water
*aquifolius -a -um*   with pointed leaves, spiny-leaved
*Aquilegia*   Eagle (claw-like nectaries)
*aquilinus -a -um*   of eagles, eagle-like (the appearance of the vasculature in the cut rhizome of *Pteridium*)
*aquilus -a -um*   blackish-brown
*arabicus -a -um, arabus -a -um*   of Arabia, Arabian
*Arabidopsis*   *Arabis*-resembler
*Arabis*   Arabian
*arachnites*   spider-like
*arachnoideus -a -um*   cobwebbed, covered with a weft of hairs
*Aralia*   origin uncertain, could be from French Canadian *aralie*

*araneosus -a -um*   spider-like, cobwebby

*aranifer -era -erum*   spider bearing

*araucanus -a -um*   from the name of a tribe of Chilean Indians

*Araucaria*   for a tribe of Chilean Indians

*arborescens*   becoming or tending to be tree-like

*arboreus -a -um*   tree-like

*arbusculus -a -um*   small tree-like, shrubby

*Arbutus*   the old Latin name

*Archangelica*   supposedly revealed to Matthaeus Sylvaticus by the archangel as a medicinal plant

*archi-*   primitive- beginning-

*arct-, arcto-*   bear-, northern-

*arcticus -a -um*   of the Artic regions, Arctic

*Arctium*   Bear (a name in Pliny, the shaggy hair)

*Arctostaphylos*   Bear's grapes (this is the Greek version of *uva-ursi*)

*Arctotis*   Bear's ear

*Arctous*   That-of-the-bear (the black bearberry)

*arcturus -a -um*   bear's tail-like

*arcuatus -a -um*   bowed, curved

*ardens*   glowing, fiery

*Aregelia*   for E. A. von Regel of St Petersburg Botanic Garden (1815–1892)

*Aremonia*   derived from the Greek plant name *Argemon* (as also is *Agrimonia*)

*Arenaria*   Sand dweller

*arenarius -a -um, arenicolus -a -um, arenosus -a -um*   growing in sand, of sandy places

*arenastrus -a -um*   resembling *Arenaria*

*areolatus -a -um*   with distinct angular spaces (in the leaves)

*Argemone*   a name used by Dioscorides for a poppy-like plant used medicinally as a remedy for cataract)

*argenteus -a -um*   silvery

*argillaceus -a -um*   growing in clay, whitish, clay-like, of clay

*argo-*   pure white-

*argutus -a -um*   sharply toothed or notched

*Argylia*   for Archibald Campbell, third Duke of Argyle and plant introducer

*argyreus -a -um, argyro-*   silvery, silver-

*aria*   a name used by Theophrastus for a whitebeam

*arianus -a -um*   from Afghanistan, Afghan

*aridus -a -um*   of dry habitats, dry, arid

*arietinus -a -um*   ram-horned

*-aris -is -e*   -pertaining to

*aristatus -a -um*   with a beard, awned, aristate (see Fig. 7(*g*))

*Aristida*   Awn (they are conspicuous)

*Aristolochia*   Birth-improver (abortifacient property)

*-arius -a -um*   -belonging to, -having

*arizelus -a -um*   notable

*arizonicus -a -um*   from Arizona, USA

*armatus -a -um*   thorny, armed

*armeniacus -a -um*   from Armenia (mistake for China), or apricot yellow

*armenus -a -um*   from Armenia, Armenian

*Armeria*   ancient Latin name for a *Dianthus*

*armillaris -is -e, armillatus -a -um*   bracelet-like of uncertain meaning (of preparation for battle?)

*arnoldianua -a -um*   of the Arnold Arboretum, Massachussets, USA

*Arnoseris*   Lamb succour

*aromaticus -a -um*   fragrant

*arrectus -a -um*   raised up, erect

*arrhen-, arrhena-*   male-, stamen-

*Arrhenatherum*   Male-awn (the lower spikelet is male and awned)

*arrhizus -a -um*   without roots, rootless

*Artemisia*   after Queen Artemisia of Caria, Asia Minor

*arthro-*   joint-, jointed-

*Arthrocnemum*   Jointed-leg

*articulatus -a -um, arthro-, arto-*   jointed, joint-, articulated

*Artocarpus*   Bread-fruit

*Arum*   a name used by Theophrastus

*arundinaceus -a -um*   reed-like

*Arundo*   the old Latin name for a reed

*arvalis -is -e*   of arable or cultivated land, from Arvas, North Spain

*arvensis -is -e*   of the field, of ploughed fields

*arvernensis -is -e*   from Auvergne, France

*Asarina*   a vernacular name for *Antirrhinum*

*Asarum*   a name used by Dioscorides

*ascendens*   upwards, ascending

*-ascens*   -becoming, -turning to

*Asclepias*   for Aesculapius, mythological god of medicine

*asco-*   bag-like-, bag-

*asininus -a -um*   ass-like (eared)

*Asparagus*   the Greek name for plants with edible young shoots *asper -era -erum*   rough (the surface texture)

*asperatus -a -um*   rough

*aspergilliformis -is -e*   shaped like a brush, with several fine branches

*aspermus -a -um*   without seed, seedless

*aspernatus -a -um*   despised

*aspersus -a -um*   sprinkled

*Asperugo*   Rough-one

*Asperula*   Little rough one

*Aspidium*   Shield (the shape of the indusium)

*Asplenium*   Spleen (its medicinal use)

*assimilis -is -e*   resembling, like, similar to

*assurgens, assurgenti*   rising upwards, ascending

*Aster*   Star

*-aster -ra -rum*   -somewhat resembling (usually implying inferiority)

*asterias*   star-like

*asterioides*   Aster-like

*asthmaticus -a -um*   of asthma (its medicinal use)

*astictus -a -um*   immaculate, without blemishes, unspotted

*Astilbe*   Without brilliance (the flowers)

*Astragalus*   a name used by Pliny for a plant with vertebra-like knotted roots

*Astrantia*   l'Ecluse's name for masterwort

*astro-*   star-shaped-

*Astrocarpus*   Star-fruit

*ater atra atrum*   matt-black

*athamanticus -a -um, athemanticus -a -um*   of Mount Athamas, Sicily

*Athanasia*   Immortal (without death, its funerary use)

*athero-*   bristle-

*Athyrium*   Sporty (from the varying structure of the sori)

*-aticus -a -um, -atilis -is -e*   -from (a place)

*atlanticus -a -um*   of the Atlas Mountains, North Africa, of Atlantic areas

*atratus -a -um*   blackish, clothed in black

*atri- atro-*   very dark-, better- (a colour)

*Atriplex*   the name used by Pliny

*Atropa*   Inflexible (one of the goddesses of fate)

*atrovirens*   very dark green

*attenuatus -a -um*   tapering

*-atus -a -um*   -having

*Aubrieta* (*Aubretia*)   for Tournefort's artist friend, Claude Aubriet (1665–1742)

*aucuparius -a -um*   bird-catching, of bird catchers (fruit used as bait, *avis capio*)

*augustus -a -um*   stately, noble, tall

*aulicus -a -um*   courtly

*aulo-*   tube-, furrowed-

*aurantiacus -a -um, aurantius -a -um*   orange-coloured

*aurarius -a -um, auratus -a -um*   golden, ornamented with gold

*aureliensis -is -e*   from Orleans, France

*aureo-, aureus -a -um*   golden-yellow

*auricomus -a -um*   with golden hair

*Auricula*   l'Ecluse's name for the bear's ear Primula, *auricula-ursi*

*auriculatus -a -um*   lobed like an ear, with ear-like lobes

*aurigeranus -a -um*   from Ariège, France

*auritus -a -um*   with ears, long-eared

*aurorius -a -um*   orange

*aurosus -a -um*   golden

*australasiacus -a -um, australiensis -is -e*   Australian, South Asiatic

*australis -is -e*   southern

*austriacus -a -um*   from Austria, Austrian

*autumnalis -is -e*   of the autumn (flowering or growing)

*avellanus -a -um*   from Avella, Italy or hazy

*Avena*   Nourishment

*avenaceus -a -um*   oat-like

*avicularis -is -e*   of small birds, eaten by small birds

*avium*   of the birds

*axillaris -is -e*   arising from the leaf axils (flowers)

*Axyris*   Without edge (the bland flavour)

*Azalea*   Of-dry-habitats (formerly used for *Loiseleuria*)

*Azolla*   thought to refer to its inability to survive out of water

*azureus -a -um*   sky-blue


*Babiana*   Baboon (they feed on the corms)

*babylonicus -a -um*   from Babylon (the willows of Ps. 137?)

*baccatus -a -um*   having berries, fruits with fleshy or pulpy coats

*baccifer -era -erum*   bearing berries

*bacillaris -is -e*   staff-like, stick-like

*badius -a -um*   reddish-brown

*baeticus -a -um*   from Spain (Baetica), Andalusian

*balansae, balansanus -a -um*   for Benedict Balansa, French plant collector (1825–1891)

*balanus*   the ancient name for an acorn

*balcanus -a -um*   of the Balkans, Balkan

*Baldellia*   for B. Bartolini-Baldelli, Italian nobleman

*baldensis -is -e*   from the area of Mount Baldo, North Italy

*baldschuanicus -a -um*   from Baldschuan, Bokhara

*Ballota*   the Greek name for one species

*balsameus -a -um*   balsam-like, yielding a balsam

*balsamifer -era -erum*   yielding a balsam, producing a fragrant resin

*banaticus -a -um*   from Banat, Romania

*Baphia*   Dye (cam-wood gives a red dye, it is also used for violin bows)

*baphicantus -a -um*   of the dyers, dyers'

*barbadensis -is -e*   from Barbados Island

*barba-jovis*   Jupiter's beard

*Barbarea*   after St Barbara

*barbarus -a -um*   foreign, from Barbary, (North African coast)

*barbatus -a -um*   with tufts of hair, bearded

*barbellatus -a -um*   having small barbs

*bargigerus -a -um*   bearded

*barometz*   from a Tartar word meaning lamb (the woolly fern's rootstock

*Barosma*   Heavy-odour

*Bartsia*   for Johann Bartsch, Dutch physician

*bary-*   heavy-

*basalis -is -e*   sessile-, basal-

*basi-*   of the base-, from the base-

*basilaris -is -e*   relating to the base

*basilicus -a -um*   princely, royal

*bastardii*   for T. Bastard, author of the flora of Maine & Loire, 1809

*babatas*   Haitian name for sweet potato

*batrachioides*   water-buttercup-like, *Batrachium*-like

*Batrachium*   Little frog (Greek for some *Ranunculus* species)

*beccabunga*   from an old German name 'Bachbungen', mouth-smart or streamlet-blocker

*Begonia*   for Michel Bégon, French Governor of Canada and patron of botany (1638–1710)

*belladonna*   beautiful lady, the juice of the deadly nightshade was used to beautify by inducing pallid skin and dilated eyes when applied as a decoction *bellatulus -a -um, bellus -a -um*   somewhat beautiful

*bellidiformis -is -e, belloides*   daisy-like, *Bellis*-like

*Bellis*   Pretty (a name used in Pliny)

*bellobatus -a -um*   beautiful bramble

*bellus -a -um*   handsome, beautiful

*benedictus -a -um*   well spoken of, blessed

*benjamina*   from an Indian name, *ben-yan*

*benzoin*   from an Arabic name

*Berberis*   Barbary (from an Arabic name for North Africa)

*bergamia*   from the Turkish name of the bergamot orange, *beg-armodi*

*berolinensis -is -e*   from Berlin, Germany

*Berteroa*   for Carlo G. L. Bertero, Italian physician (1789–1831)

*Berula*   the Latin name in Marcellus Empyricus

*Beta*   the Latin name for beet

*betaceus -a -um*   beet-like, resembling *Beta*

*Betonica*   from a name in Pliny for a medicinal plant from Vettones, Spain

*betonicifolius -a -um*   betony-leaved

*Betula*   Pitch (the name in Pliny, bitumen is distilled from the bark)

*betulinus -a -um, betuloides, betulus -a -um*   birch-like

*bi-, bis-*   two-, twice-

*bicapsularis -is -e*   having two capsules

*bicolor*   of two colours

*Bidens*   Two teeth (the scales at the fruit apex)

*biennis -is -e*   lasting for two years, biennial

*bifarius -a -um*   in two opposed ranks (leaves or flowers)

*bifidus -a -um*   deeply two-cleft, bifid

*bijugus -a -um*   two-together, yoked

*-bilis -is -e*   -able, -capable

*Billardiera, billardierei (billardierii)*   for J. J. H. de la Billardière, French botanist (1755–1834)

*binatus -a -um*   with two leaflets, bifoliate

*Biophytum*   Life plant (sensitive leaves)

*Biscutella*   Two trays (the form of the fruit)

*biserratus -a -um*   twice-toothed, double-toothed (leaf margin teeth themselves toothed)

*bistortus -a -um*   twice-twisted (the roots, from the medieval name for bistort)

*bithynicus -a -um*   from Bithynia, Asia Minor

*bituminosus -a -um*   tarry, clammy, adhesive

*Bixa*   from a South American native name for the annatto tree

*Blackstonia*   for John Blackstone, English botanical writer
*blandus -a -um*   pleasing, charming, not harsh, bland
*blattarius -a -um*   cockroach-like, an ancient Latin name
*Blechnum*   the Greek name for a fern
*blepharo-*   fringe-, eyelash-
*blepharophyllus -a -um*   with fringed leaves
*Blumenbachia*   for Johann Friedrich Blumenbach of Göttingen, medical doctor (1752–1840)
*Blysmus*   meaning uncertain
*boeoticus -a -um*   from Boeotia, near Athens, Greece
*bolanderi*   for Prof. H. N. Bolander of Geneva, plant collector in California and Oregon, USA (1831–1897)
*Bombax*   Silk (the hair covering of the seeds)
*bombyci-*   silk- (Greek *bombyx*)
*bombycinus -a -um*   silky
*bona-nox*   good night (night-flowering)
*bonariensis -is -e*   from Buenos Aires, Argentina
*bondus*   an Arabic name for a hazel-nut
*bononiensis -is -e*   from Bologna, North Italy
*bonus-henricus*   good King Henry (allgood or mercury)
*Boophone*   Ox-killer (narcotic property)
*Borago*   possibly Shaggy coat (the leaves)
*borbonicus -a -um*   from Reunion Island, Indian Ocean, or for the French Bourbon Kings
*borealis -is -e*   northern
*Boreava*   for Alexander Boreau, Belgian botanist (1803–1875)
*Boronia*   for Francesco Boroni, assistant to Humphrey Sibthorp in Greece
*Borreri*   for W. Borrer, British botanist 1781–1862
*bothrio-*   minutely pitted-
*botry-*   bunched-, panicled-
*Botrychium*   Little-bunch (the fertile portion of the frond of moonwort)
*botryoides, botrys*   resembling a bunch of grapes
*botrytis -is -e*   racemose, racemed
*botulinus -a -um*   shaped like small sausages (branch segments)
*Bougainvillea*   for Louis Antoine de Bougainville, French navigator (1729–1811)
*brachiatus -a -um*   arm-like, branched at about a right-angle
*brachy-*   short-
*brachybotrys*   short-clustered, shortly bunched
*Brachychiton*   Short-tunic
*Brachypodium*   Short-foot

*Brachystelma*   Short-crown (the coronna)

*bracteatus -a -um*   with bracts, bracteate (as in the inflorescences of *Hydrangea, Poinsettia* and *Acanthus*)

*bracteosus -a -um*   with large or conspicuous bracts

*brandisianus -a -um, brandisii*   for Sir Dietrich Brandis, dendrologist of Bonn (1824–1907)

*brasiliensis -is -e*   from Brazil, Brazilian

*Brassica*   Pliny's name for various cabbage-like plants

*brassici-*   cabbage-, *Brassica-*

*brevi-, brevis -is -e*   short-, abbreviated-

*breviscapus -a -um*   short-stalked, with a short scape

*Briza*   Food grain (an ancient Greek name for rye)

*Bromus*   Food (the Greek name for an edible grass)

*-bromus*   -smelling, -stinking

*bronchialis -is -e*   throated, of the lungs (medicinal use)

*brumalis -is -e*   of the winter solstice, winter-flowering

*Brunfelsia*   for Otto Brunfels (1489–1534) who made the first critical plant drawings

*brunneus -a -um*   russet-brown

*Bryanthus*   Moss-flower

*bryoides*   moss-like

*Bryonia*   Sprouter (a name used by Dioscorides)

*bubalinus -a -um*   of cattle, of oxen

*buccinatorius -a -um, buccinatus -a -um*   heralded, trumpet-shaped, horn-shaped

*bucephalus -a -um*   bull-headed

*Buchanani*   either for Francis Buchanan Hamilton of Calcutta Botanic Garden or for John Buchanan, specialist on New Zealand plants

*bucharicus -a -um*   from Bokhara, Turkistan

*bucinalis -is -e, bucinatus -a -um*   trumpet-shaped, trumpet-like

*Buda*   an Adansonian name of no meaning

*Buddleia (Buddleja)*   for Adam Buddle, English 17th-century botanist

*bufonius -a -um*   of the toad, living in damp places

*bulbifer -era -erum*   producing bulbs (often when these take the place of normal flowers)

*bulbi-, bulbo-*   bulb-, bulbous-

*Bulbine*   the Greek name for a bulb

*bulbocastanus -a -um*   chestnut-brown-bulbed

*bulbosus -a -um*   swollen, having bulbs, bulbous

*bullatus -a -um*   with a bumpy surface, puckered, blistered, bullate *bumannus -a -um*   having large tubercles

*-bundus -a -um*   -having the capacity for

*Bunias*   the Greek name for a kind of turnip

*Bunium*   a name used by Dioscorides

*buphthalmoides*   ox-eyed

*Bupleurum*   Ox-rib, a name used by Nicander

*burmanicus -a -um*   from Burma, Burmese

*bursa-pastoris*   shepherd's purse

*Butomus*   Ox-cutter, a name used by Theophrastus with reference to the sharp-edged leaves

*butyraceus -a -um*   oily, buttery

*Butyrospermum*   Butter seed (oily seed of shea-butter tree)

*buxifolius -a -um*   box-leaved

*Buxus*   a name used by Virgil for *B. sempervirens*

*byzantinus -a -um*   from Istanbul (Byzantium), Turkish


*cacao*   Aztec name for the cocoa tree, *Theobroma*

*cachemerianus -a -um, cachemiricus -a -um*   from Kashmir

*cacti-*   cactus-like- (originally the Greek cactus was an Old World spiny plant, not one of the *Cactaceae*)

*cacumenus -a -um*   of the mountain top

*cadmicus -a -um*   with a metallic appearance

*caducus -a -um*   transient, not persisting, caducous

*caeno- caenos-*   fresh-, recent-

*caenosus -a -um*   muddy, growing on mud

*caerulescens*   turning blue, bluish

*caeruleus -a -um*   dark sky-blue

*caesius -a -um*   bluish-grey, lavender-coloured

*caespitosus -a -um*   growing in tufts, matted, tussock-forming

*caffer -ra -rum, caffrorum*   from South Africa, of the unbelievers (Kaffirs)

*cainito*   the West Indian name for the star apple

*Caiophora*   Burn-carrier (the stinging hairs)

*cairicus -a -um*   from Cairo, Egypt

*cajan*   the Malay name for pigeon pea

*Cajanus*   the Malay name

*cajennensis -is -e*   from Cayenne, French Guiana

*cajuputi*   the Malayan name

*Cakile*   from an Arabic name

*cala-*   beautiful-

*calanthus -a -um*   beautiful-flowered

*calaba*   the West Indian name

*calabricus -a -um*   from Calabria, Italy

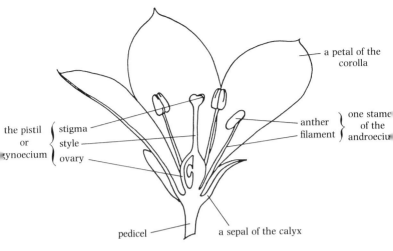

Fig. 1. The parts of a flower as seen in a stylized flower which is cut vertically in half.

*Calamagrostis*   Reed grass, name used by Dioscorides
*calamarius -a -um*   reed-like, resembling *Calamus*
*calaminaris -is -e*   cadmium-red, growing on the zinc ore, calamine
*Calamintha*   beautiful mint
*calamitosus -a -um*   causing loss, dangerous
*Calanthe*   Beautiful flower
*calathinus -a -um*   basket-shaped, basket-like
*calcaratus -a -um*, *calcatus -a -um*   spurred, having a spur
*calcareus -a -um*   of lime-rich soils
*calceolatus -a -um*   shoe-shaped, slipper-shaped
*calceolus -a -um*   like a small shoe
*calcicolus -a -um*   living on limy soils
*calcifugus -a -um*   disliking lime, avoiding limy soils
*calcitrapa*   caltrop (the fruit's resemblance to the spiked ball used to damage the hooves of charging cavalry horses)
*caledonicus -a -um*   from Scotland (Caledonia), Scottish, of northern Britain
*Calendula*   First day of the month (Latin *calendae*) associated with paying accounts and settling debts)
*Calepina*   an Adansonian name perhaps relating to Aleppo

*calidus -a -um*   fiery, warm

*californicus -a -um*   from California, USA

*caliginosus -a -um*   of misty places

*Calla*   Beauty (a name used in Pliny)

*calli-, callis-*   beautiful-

*Calliandra*   Beautiful stamens (shaving-brush tree)

*callifolius -a -um*   *Calla*-leaved

*callimorphus -a -um*   of beautiful form or shape

*Callistemon*   Beautiful stamens (bottle-brush tree)

*Callistephus*   Beautiful crown

*callistus -a -um*   very beautiful

*Callitriche*   Beautiful hair

*callosus -a -um*   hardened

*Calluna*   Brush (former common use for sweeping)

*calo-*   beautiful-

*calomelanos*   beautifully-dark

*calophrys*   with dark margins

*calostomus -a -um*   beautiful mattress (growth habit)

*calpophilus -a -um*   estuary-loving, estuarine

*Caltha*   old Latin name, probably for marigold

*calvescens*   with non-persistent hair, becoming bald

*calvus -a -um*   naked, hairless, bald

*calyc-, calyci-*   calyx-

*calycinus -a -um, calycosus -a -um*   with a persistent calyx, calyx-like

*calyculatus -a -um*   resembling a small calyx

*calyptr- calyptro-*   hooded-, lidded-

*calyptratus -a -um*   with a cap-like cover over the flowers or fruits

*Calystegia*   Calyx-cover (the calyx is at first obscured by prophylls)

*camaldulensis -is -e*   from the Camaldoli gardens near Naples

*camara*   a West Indian name, arched

*Camassia*   a North American Indian name for an edible bulb (Quamash)

*cambodgensis -is -e*   from Cambridge

*cambodiensis -is -e*   from Cambodia, South East Asia

*cambrensis -is -e, cambricus -a -um*   from Wales (Cambria), Welsh

*Camelina*   Little-flax

*Camellia*   for George J. Kamel, Asiatic traveller (1661–1706)

*camelliiflorus -a -um*   *Camellia*-flowered

*cammarus -a -um*   lobster (from a name used by Dioscorides)

*campani-, campanularius -a -um, campanulatus -a -um, campanulus -a -um*   bell-shaped, bell-flower-like
*Campanula*   Little bell
*campester -tris -tre*   of the pasture, from flat land
*camphoratus -a -um*   camphor-like scented
*campto-*   bent-
*campyl-, campylo-*   bent-, curved-
*camtschatcensis -is -e, camtschaticus -a -um*   from the Kamchatka Peninsula, Siberia
*canadensis -is -e*   from Canada, Canadian
*canaliculatus -a -um*   furrowed, channelled
*cananga*   from a Malayan name
*canariensis -is -e*   from the Canary Isles, of bird food
*canarinus -a -um*   yellowish, resembling *Canarium*
*canarius -a -um*   canary-yellow
*cancellatus -a -um*   cross-banded, chequered
*candelabrum*   candle-tree, like a branched candlestick
*candicans*   whitish, hoary-white, with white woolly hair
*candidus -a -um*   shining-white
*canephorus -a -um*   like a basket bearer
*canescens*   turning hoary-white, canescent
*caninus -a -um*   of the dog, sharp-toothed or spined, wild or inferior
*cannabinus -a -um*   hemp-like, resembling *Cannabis*
*Cannabis*   Dioscorides' name for hemp
*cano-*   hairy-
*cantabricus -a -um*   from Cantabria, North Spain
*cantabrigiensis -is -e*   from Cambridge (Cantabrigia)
*cantianus -a -um*   from Kent, England
*canus -a -um*   whitish-grey
*capensis -is -e*   from Cape Colony, South Africa
*capillaceus -a -um, capillaris -is -e, capillatus -a -um*   hair-like, very slender
*capillipes*   with a very slender stalk
*capillus-veneris*   Venus' hair
*capitatus -a -um*   growing in a head, head-like (inflorescence)
*capitulatus -a -um*   having small heads
*capnoides*   smoke-like
*cappadocicus -a -um, cappadocius -a -um*   from Cappadocia, Asia Minor
*capraeus -a -um*   of the goat, goat-like (smell) *capraea* = she-goat
*capreolatus -a -um*   tendrilled, with tendrils, twining
*Caprifolium*   Goat-leaf

*Capriola*   Goat
*Capsella*   Little box (the form of the fruit)
*Capsicum*   Biter (the hot taste)
*capsularis -is -e*   producing capsules
*caput-medusae*   Medusa's head
*caracalla*   beautiful snail, cloaked
*caracasanus -a -um*   from Caracas
*carataviensis -is -e*   from Karatau, Kazakhstan, USSR
*Cardamine*   Dioscorides' name for cress
*Cardaminopsis*   *Cardamine*-resembler
*cardamomum*   ancient Greek name for the Indian spice
*Cardaria*   Heart-like (the fruiting pods)
*cardi-, cardio-*   heart-shaped-
*cardiacus -a -um*   of heart conditions (medicinal use)
*cardinalis -is -e*   cardinal-red
*cardiopetalus -a -um*   with heart-shaped petals
*cardunculus -a -um*   thistle-like
*Carduus*   Thistle (a name in Virgil)
*Carex*   Cutter (the sharp leaf margins of many)
*caribaeus -a -um*   from the Caribbean
*caricinus -a -um, caricosus -a -um*   sedge-like, resembling *Carex*
*caricus -a -um*   from Caria, province of Asia Minor
*carinatus -a -um*   keeled, having a keel-like ridge
*Carlina*   for Charlemagne
*carmineus -a -um*   carmine
*Carnegiea*   for the philanthropist Andrew Carnegie
*carneus -a -um, carnicolor*   flesh-coloured
*carniolicus -a -um*   from Carniola, Yugoslavia
*carnosus -a -um*   fleshy, thick and soft textured
*carolinianus -a -um, carolinus -a -um*   of North or South Carolina, USA
*carota*   the old name for carrot (*Daucus carota*)
*carpathicus -a -um, carpaticus -a -um*   from the Carpathian Mountains
*carpetanus -a -um*   from the Toledo area of Spain
*carpini-*   hornbeam-like-
*Carpinus*   ancient Latin name for hornbeam, some derive it from Celtic for a yoke
*carpo-, carpos-, -carpus -a -um*   fruit-, -fruited
*Carpobrotus*   Edible fruit
*Carpodetus*   Bound-fruit (external appearance of the Putaputawheta fruit)

*Carrichtera*   for Bartholomaeus Carrichter, physician to
  Maximillian II
*Carthamus*   Painted-one (Hebrew *qarthami*, an orange-red dye is
  made from *C. tinctorius*)
*carthusianorum*   of the Grande Chartreuse Monastery of
  Carthusian Monks, Grenoble, France
*cartilagineus -a -um*   cartilage-like (texture of some part, e.g. leaf
  margin)
*Carum*   Dioscorides' name for caraway
*carunculatus -a -um*   with a prominent caruncle (outgrowth of
  the seed coat, usually obscuring the micropyle)
*carvi (carui)*   from Caria, Asia Minor
*Carya*   ancient Greek name for a walnut
*caryo-*   nut-, clove-
*Caryolopha*   Crest-of-nuts (they form a ring)
*caryophylleus -a -um*   resembling a stitchwort, clove-coloured
*cashemirianus -a -um*   see *cachemerianus*
*caspicus -a -um*   of the Caspian area
*Cassia*   a name used by Dioscorides from a Hebrew name
  (*quetsi'oth*) used by Linnaeus for *C. fistula* (medicinal senna)
*cassioides*   resembling *Cassia*
*cassubicus -a -um*   from Cassubia, part of Pomerania
*Castalia*   of the spring of the Muses on Mount Parnassus
*Castanea*   old Latin name for the sweet chestnut
*castaneus -a -um, castanus -a -um*   chestnut-brown
*castello-paivae*   for Baron Castello de Paiva
*castus -a -um*   spotless, pure
*cat-, cata-, cato-*   below-, outwards-, downwards-, from-, under-,
  against-, along-
*Catabrosa*   Eaten (the appearance of the tip of the lemmas)
*catacosmus -a -um*   adorned
*catafractus -a -um, cataphractus -a -um*   enclosed, armoured,
  closed in
*catalpa*   an East Indian name
*Catananche*   Driving-force (its use in love potions by Greek
  women)
*Catapodium*   Minute-stalk (the spikelets are subsessile)
*catappa*   from a native East Indian name for olive-bark tree
*cataria*   of cats, old name for catmint (catnip)
*catarractae, catarractarum*   growing near waterfalls, resembling a
  waterfall
*catenarius -a -um, catenatus -a -um*   chain-like

*catharticus -a -um*   purgative, purging, cathartic

*cathayanus -a -um, cathayensis -is -e*   from China (Cathay)

*catholicus -a -um*   world-wide, universal, of Catholic lands (Spain and Portugal)

*Cattleya*   for Wm. Cattley, English plant collector, and patron of Botany

*Caucalis*   old Greek name for an Umbelliferous plant

*caucasicus -a -um*   from the Caucasus

*caudatus -a -um, caudi-*   tailed (see Fig. 7(a))

*caudiculatus -a -um*   with a thread-like caudicle or tail

*caulescens*   having a distinct stem

*cauliatus -a -um, -caulis -is -e, -caulo, -caulos*   of the stem or stalk, -stemmed, -stalked

*cauliflorus -a -um*   bearing flowers on the main stem, flowering on the old woody stem

*causticus -a -um*   with a caustic taste (mouth-burning)

*cauticolus -a -um*   growing on cliffs, cliff-dwelling

*cautleoides*   resembling *Cautlea*

*cavernicolus -a -um*   growing in caves, cave-dwelling

*cavernosus -a -um*   full of holes

*cavus -a -um*   hollow

*cayennensis -is -e*   from Cayenne, French Guiana

*Cecropia*   for Cecrops, legendary King of Athens

*Cedrus*   the Greek name for a Juniper

*Ceiba*   from a South American name

*Celastrus*   Theophrastus' name for an evergreen tree

*celebicus -a -um*   from the Indonesian Island of Celebes

*celeratus -a -um*   hastened

*-cellus -a -um*   -lesser, -somewhat

*Celosia*   Burning (flower colour)

*celtibiricus -a -um*   from central Spain

*Celtis*   Linnaeus applied this old Greek name to the European hackberry

*cembra*   the old name for the arolla or stone pine

*cembroides, cembrus -a -um*   resembling *Pinus cembra*

*cenisius -a -um*   from Mt Cenis on the French–Italian border

*ceno-, cenose-*   empty-

*Centaurea*   Centaur (mythical creature with the body of a horse replacing the hips and legs of a man, the name used by Hippocrates)

*centaureoides*   resembling *Centaurea*

*Centaurium*   for the Centaur, Chiron, who was fabled to have used this plant medicinally

*centi-*   one hundred-, many-

*centra-, centro-, -centrus -a -um*   spur-, -spurred

*centralis -is -e*   in the middle, central

*Centranthus (Kentranthus)*   Spur-flower

*cepa*   the old name for an onion

*cepaeus -a -um*   growing in gardens, from the ancient Greek for
   a salad plant

*cephal-*   head, head-like-

*Cephalanthera*   Head anther (its position on the column)

*Cephalaria*   Head (the capitate inflorescence)

*cephalidus -a -um*   having a head

*cephalonicus -a -um*   from Cephalonia, one of the Ionian Islands

*cephalotes*   having a small head-like appearance

*cephalotus -a -um*   with flowers in a large head

*-cephalus -a -um*   -headed

*-ceras*   -horned

*ceraseus -a -um*   waxy

*cerasifer -era -erum*   bearing cherries (cherry-like fruits)

*cerasinus -a -um*   cherry-red

*Cerastium*   Horned (the fruiting capsule's shape)

*Cerasus*   from an Asiatic name for the sour cherry

*cerato-*   horn-shaped-

*Ceratochloa*   Horned-grass (the lemmas are horn-like)

*Ceratophyllum*   Horn leaf (the texture of the leaf)

*Ceratopteris*   Horned fern

*cerealis -is -e*   for Ceres, the goddess of agriculture

*cereus -a -um*   waxy (*cereus* = wax taper)

*cerifer -era -erum*   wax-bearing

*cerinus -a -um*   waxy

*cernuus -a -um*   drooping, curving forwards

*Ceropegia*   Fountain of wax (appearance of the inflorescence)

*Ceroxylon*   Wax-wood

*cerris*   the ancient Latin name for turkey oak

*cervicarius -a -um*   constricted, keeled

*cervinus -a -um*   tawny, stag-coloured

*cespitosus -a -um*   see *caespitosus*

*Cestrum*   an ancient Greek name

*Ceterach*   an Arabic name for a fern 'chetrak'

*cevisius -a -um*   closely resembling

*ceylanicus -a -um*   from Ceylon

*chaeno-*   splitting-, gaping-

*Chaenomeles*   Gaping apple

*Chaenorrhinum*   Gaping nose (analogy with *Antirrhinum*)

79

*chaero-* pleasing-, rejoicing-

*Chaerophyllum* Pleasing leaf (the ornamental foliage)

*chaeto-* long hair-like-

*chalcedonicus -a -um* from Chalcedonia, Turkish Bosphorus

*chamae-* on-the-ground-, ground-hugging-, lowly-, low-growing-, prostrate-

*Chamaecyparis* Dwarf-cypress

*chamaedrys* ground oak

*Chamaemelum* Ground-apple (the fragrance and habit)

*Chamaenerion* Dwarf-oleander

*Chamaepericlymenum* Dwarf-climbing-plant

*chamaeunus -a -um* lying on the ground

*characias* the name in Pliny for a spurge with very caustic latex

*charantius -a -um* graceful

*charianthus -a -um* with elegant flowers

*Charieis* Elegant

*chartaceus -a -um* parchment-like

*chasmanthus -a -um* having open flowers

*chauno-* gaping-

*Cheilanthes* Lip-flower (the infolded edge)

*cheilanthus -a -um* with lipped flowers

*cheilo-* lip-, lipped-

*Cheiranthus* Red-flower (from an Arabic name)

*cheiri* red-flowered (from an Arabic name)

*cheiro-* hand-, hand-like-

*Chelidonium* Swallow (flowering at the time of their arrival)

*Chenopodium* Goose-foot (the shape of the leaves)

*cherimola* a Peruvian-Spanish name

*Cherleria* for J. H. Cherler, son-in-law of C. H. Bauhin

*chermisinus -a -um* red

*chia* from the Greek Island of Chios

*chilensis -is -e* from Chile, Chilean

*chiloensis -is -e* from Chiloe Island off Chile

*-chilus -a -um* -lipped

*chima-, chimon-* winter-

*chimaera* monstrous, fanciful

*Chimonanthus* Winter flower

*chinensis -is -e* from China, Chinese

*chio-, chion-, chiono-* snow-

*Chionodoxa* Glory of the snow (very early flowering)

*Chironia, chironius -a -um* after Chiron, the centaur of Greek mythology who taught the medicinal use of plants

*chirophyllus -a -um* with hand-shaped leaves

*chlamy-*   cloak- cloaked-

*chlor-, chloro-, chlorus -a -um*   yellowish-green-

*Chlora*   Greenish-yellow-one

*chloranthus -a -um*   green-flowered

*Chloris*   for Chloris, Greek goddess of flowers

*chlorophyllus -a -um*   green-leaved

*chocolatinus -a -um*   chocolate-brown

*chondro-*   rough-, angular, lumpy, coarse-

*chordatus -a -um*   cord-like

*chordo-*   string-, slender-elongate-

*chori-*   separate-, apart-

*Chorispora*   Separated-seed (winged seeds are separated within
  the fruit)

*-chromatus -a -um, -chromus -a -um*   -coloured

*chrono-*   time-

*Chrysanthemum*   Golden-flower (Dioscorides' name for
  C. coronarium)

*chrysanthus -a -um*   yellow-flowered

*chryseus -a -um, chrys-, chryso-*   golden-yellow

*Chrysocoma*   Golden-hair (the terminal inflorescence)

*chrysographes*   marked with gold lines, as if written upon in
  gold

*chrysomallus -a -um*   with golden wool, golden-woolly-hairy

*chrysops*   with a golden eye

*Chrysosplenium*   Golden-spleenwort (used for diseases of the
  spleen)

*chrysostomus -a -um*   with a golden throat

*chrysotoxus -a -um*   golden-arched

*chyllus -a -um*   from a Himalayan vernacular name

*chylo-*   sappy-

*cibarius -a -um*   edible

*cicatricatus -a -um*   marked with scars (left by falling structures
  such as leaves)

*Cicenda*   an Adansonian name with no obvious meaning

*cicer*   the old Latin name for the chick-pea

*Cicerbita*   an old Latin name for a thistle

*Cichorium*   from an Arabic name

*ciconius -a -um*   resembling a stork's neck

*Cicuta*   the Latin name for hemlock

*cicutarius -a -um*   resembling Cicuta, with large two- or three-
  pinnate leaves

*ciliaris -is -e, ciliatus -a -um, ciliosus -a -um*   fringed with hairs,
  ciliate

*cilicicus -a -um*   from Cilicia, southern Turkey

*-cillus -a -um*   -lesser

*cincinnatus -a -um*   with crisped hairs

*cinctus -a -um*   girdled

*cineraceus -a -um, cinerarius -a -um, cinerascens*   ash-coloured, covered with ash-grey felted hairs

*Cineraria*   Ashen-one (the foliage colour)

*cinereus -a -um*   ash-grey

*cinnabarinus -a -um*   cinnabar-red

*cinnamomeus -a -um*   cinnamon-brown

*Cinnomomum*   the Greek name used by Theophrastus

*Circaea*   for the enchantress Circe of mythology (Pliny's name for a charm plant)

*circinalis -is -e, circinatus -a -um*   curled round, coiled like a crozier, circinate

*circum-*   around-

*cirratus -a -um, cirrhatus -a -um, cirrhiferus -a -um*   having or carrying tendrils

*cirrhosus -a -um*   with large tendrils

*Cirsium*   the ancient Greek name for a thistle

*Cissus*   the ancient Greek name for ivy

*Cistus*   Capsule (conspicuous in fruit)

*citratus -a -um*   Citrus-like

*citreus -a -um, citrinus -a -um*   citron-yellow

*citri-*   citron-like-

*citriodorus -a -um*   citron-scented, lemon-scented

*Citrulus*   Little-orange (the fruit colour)

*Cladium*   Small branch

*clado-*   shoot-, branch-, of the branch-

*clandestinus -a -um*   concealed, hidden, secret

*clarus -a -um*   clear

*clausus -a -um*   shut, closed

*clavatus -a -um, clavi-, clavus -a -um*   clubbed, club-shaped

*claviculatus -a -um*   having tendrils, tendrilled

*clavigerus -a -um*   club-bearing

*Claytonia*   for John Clayton, British botanist in America (1686–1773)

*cleio- cleisto-*   shut-, closed-

*Clematis*   the Greek name for several climbing plants

*clematitis -is -e*   vine-like, with long vine-like twiggy branches

*Clianthus*   Glory-flower (the glory pea)

*clino-*   prostrate-, bed-

*Clinopodium*   Bed-foot (Dioscorides' name for the shape of the inflorescence) *clipeatus -a -um*   shield-shaped

*clivorum*   of the hills

*Clutia* (*Cluytia*)   for Outgers Cluyt (Clutius) of Leyden (1590–1650)

*clymenus -a -um*   for an ancient Greek name

*clypeatus -a -um, clypeolus -a -um*   like a Roman shield

*Clypeola* (*Clipeola*)   Shield (the shape of the fruit)

*cneorum*   of garlands, the Greek name for an olive-like shrub

*Cnicus*   the Greek name of a thistle used in dyeing

*co-, col-, con-*   together-, together with-, firmly-

*coacervatus -a -um*   clustered, in clumps

*coaetaneus -a -um*   ageing together (leaves and flowers both senesce together)

*coagulans*   curdling

*coarctatus -a -um*   pressed together, bunched, contracted

*coca*   the name used by South American Indians

*cocciferus -a -um, coccigerus -a -um*   bearing berries

*coccineus -a -um*   scarlet (the dye produced from galls on *Quercus coccifera*)

*-coccus -a -um*   -berried

*Cochlearia*   Spoon (the shape of the basal leaves)

*cochlearis -is -e*   spoon-shaped

*cochleatus -a -um*   twisted like a snail-shell, cochleate

*cochlio- cochlo-*   spiral-, twisted-

*Cocos*   from the Portuguese for monkey (*coco*)

*-codon*   -bell, -mouth

*Codonopsis*   Bell-like (flower shape)

*coelestinus -a -um, coelestis -is -e, coelestus -a -um, coeli-*   sky-blue, heavenly

*coeli-rosa*   rose of heaven

*coelo-*   hollow-

*Coeloglossum*   Hollow tongue (the lip of the flower)

*coen-, coenos-*   common-

*coerulescens*   bluish

*coeruleus -a -um*   blue

*Coffea*   from the Arabic name

*coggygria*   the ancient Greek name for *Cotinus*

*cognatus -a -um*   closely related to

*Coix*   the ancient Greek name for Job's tears grass

*Cola*   from the West African name

*Colchicum*   Dioscorides' name for *C. speciosum*

*colchicus -a -um*   from Colchis, the Caucasian area once famous for poisons

*coleatus -a -um*   sheath-like

*coleo-*   sheath-

*Coleus*   Sheath (the filaments around the style)

*collinus -a -um*   of the hills, growing on hills

*-collis -is -e*   -necked

*colocynthis*   ancient Greek name for *Citrullus colocynthis*

*colombinus -a -um*   dove-like

*colonus -a -um*   forming a mound, humped

*colorans, coloratus -a -um, -color*   coloured

*colubrinus -a -um*   snake-like

*columbarius -a -um, columbrinus -a -um*   dove-like, dove-coloured, of doves

*columnaris -is -e*   pillar-like, columnar

*Columnea*   for Fabio Colonna of Naples, publisher of *Phytobasanos* (1567–1650)

*-colus -a -um*   -loving, -inhabiting, -dwelling (follows a place or habitat)

*Colutea*   a name used by Theophrastus for a tree

*com-*   with-, together with-

*comans, comatus -a -um*   hairy-tufted, hair-like

*Comarum*   from Theophrastus' name for the strawberry tree (their similar fruits)

*Commelina*   for Caspar Commelijn, Dutch botanist (1667–1731)

*commixtus -a -um*   mixed together, mixed up

*communis -is -e*   growing in clumps, gregarious, common

*commutatus -a -um*   changed, altered (e.g. from previous inclusion in another species)

*comorensis -is -e*   from Comoro Islands, off Mozambique, East Africa

*comosus -a -um*   shaggy-tufted, with tufts formed from hairs or leaves or flowers, long-haired

*compactus -a -um*   close-growing, closely packed together

*complanatus -a -um*   flattened out upon the ground

*complexus -a -um*   encircled, embraced

*compositus -a -um*   with flowers in a head, *Aster*-flowered, compound

*compressus -a -um*   flattened sideways (as in stems), pressed together

*comptus -a -um*   ornamented, with a headdress

*con-*   with-, together with-

*concatenans, concatenatus -a -um*   joined together, forming a chain

*concavus -a -um*   basin-shaped, concave
*conchae-*, *conchi-*   shell-, shell-like-
*conchifolius -a -um*   with shell-shaped leaves
*concinnus -a -um*   well-proportioned, neat, elegant
*concolor*   uniformly-coloured, coloured similarly
*condensatus -a -um*   crowded together
*condylodes*   knobbly, with knuckle-like bumps
*confertus -a -um*   crowded, pressed together
*conformis -is -e*   symmetrical, conforming to type or relationship
*confusus -a -um*   easily mistaken for another species, intricate
*congestus -a -um*   arranged very close together
*conglomeratus -a -um*   clustered, crowded together
*conicus -a -um*   cone-shaped, conical
*conii-*   hemlock-like, resembling *Conium*
*Conium*   the Greek name for hemlock plant and poison
*conjugalis -is -e*, *conjugatus -a -um*   joined together in pairs,
   conjugate
*conjunctus -a -um*   joined together
*connatus -a -um*   united at the base
*connivens*   converging, connivent
*cono-*   cone-shaped-
*conoideus -a -um*   cone-like
*Conopodium*   Cone foot
*conopseus -a -um*   cloudy, gnat-like
*Conringia*   for Hermann Conring, German academic
*consanguineus -a -um*   closely related, of the same blood
*consimilis -is -e*   much resembling
*Consolida*   Make firm (the ancient Latin name from its use in
   healing medicines)
*consolidus -a -um*   stable, firm
*conspersus -a -um*   speckled, scattered
*conspicuus -a -um*   easily seen, marked, conspicuous
*constrictus -a -um*   erect, dense
*contemptus -a -um*   despising, despised
*contiguus -a -um*   close and touching, closely related
*contorus -a -um*   twisted, bent
*contra-*, *contro-*   against-
*contractus -a -um*   drawn together
*controversus -a -um*   doubtful, controversial
*Convallaria*   Valley (the natural habitat of lily-of-the-valley)
*convalliodorus -a -um*   lily-of-the-valley scented
*conversus -a -um*   turning towards, turning together
*convexus -a -um*   humped, bulged outwards, convex

*convolutus -a -um*   rolled together

*Convolvulus*   Interwoven (a name in Pliny)

*Conyza*   a name used by Theophrastus

*copallinus -a -um*   from a Mexican name, yielding copal-gum

*copiosus -a -um*   abundant, copious

*copticus -a -um*   from Coptos, near Thebes, Egyptian

*coracinus -a -um*   raven-black

*coraensis -is -e*   from Korea, Korean

*corallinus -a -um, corallioides*   coral-red, coral-like

*Corallorhiza*   Coral-root (the rhizomes)

*corbularia*   like a small basket

*Corchorus*   the Greek name for jute

*cordatus -a -um, cordi-*   heart-shaped, cordate (see Fig. 6(e))

*cordifolius -a -um*   with heart-shaped leaves

*Cordyline*   Club (some have large club-shaped roots)

*coreanus -a -um*   from Korea, Korean

*Coreopsis*   Bug-like (the shape of the fruits)

*coriaceus -a -um*   tough, leathery, thick-leaved

*Coriandrum*   Theophrastus' name for *C. sativum*

*Coriaria*   Leather (used in tanning)

*coriarius -a -um*   of tanning, leather-like

*corid-*   *Coris*-like

*corii-*   leathery-

*coritanus -a -um*   resembling *Coris*, from the East Midlands (home of the Coritani tribe of ancient Britons)

*corneus -a -um*   horny

*corni-, cornifer -era -erum, corniger -era -erum, -cornis -is -e*   horned, horn-bearing

*corniculatus -a -um*   having horn- or spur-like appendages or structures

*cornubiensis -is -e*   from Cornwall (Cornubia), Cornish

*cornucopiae*   horn of plenty, horn full

*Cornus*   the Latin name for the cornelian cherry

*cornutus -a -um*   horn-shaped, horned

*corollinus -a -um*   with a conspicuous corolla

*coronarius -a -um*   garlanding, forming a crown

*coronatus -a -um*   crowned

*Coronilla*   Little crown

*Coronopus*   Theophrastus' name for crowfoot (leaf-shape)

*Corrigiola*   Shoe-thong (the slender stems)

*corrugatus -a -um*   wrinkled, corrugated

*corsicus -a -um*   from Corsica, Corsican

*Cortaderia*   Cutter (from the Spanish-American name which refers to the sharp-edged leaves)

*corticalis -is -e, corticosus -a -um*   with a notable or pronounced bark

*coryandrus -a -um*   with helmet-shaped stamens

*Corydalis*   Crested lark (the spur of the flowers)

*corylinus -a -um, coryli-, Corylopsis*   hazel-like, resembling *Corylus*

*Corylus*   Helmet (the Latin name for the hazel)

*corymbosus -a -um*   with flowers arranged in corymbs, with a flat-topped raceme (see Fig. 2(*d*))

*coryne- coryno-*   club-, club-like-

*Corynephorus*   Club-bearer (the awns)

*corynephorus -a -um*   clubbed, bearing a club

*coryph-*   at the summit-

*corys-, -corythis -is -e*   helmet-, -cucculate (Greek *koris*)

*Cosmos*   Beautiful

*costalis -is -e, costatus -a -um*   with prominent ribs, with a prominent mid-rib

*Cotinus*   ancient Greek name for a wild olive

*Cotoneaster*   Quince-like

*Cotula*   Small cup

*Cotyledon*   Cupped (the leaf shape)

*coum*   from a Hebrew name

*cous*   Coan, from the island of Cos, Turkey

*cracca*   ancient Greek name for a vetch

*Crambe*   ancient Greek name for a cabbage-like plant

*crassi-*   thick-, fleshy-

*crassicaulis -is -e*   thick-stemmed

*Crassula*   Succulent-little-plant

*crassus -a -um*   thick, fleshy

*Crataegus*   Strong (the name used by Theophrastus)

*crateri-, cratero-*   strong-, goblet-shaped-

*creber -ra -rum, crebri-*   densely clustered, frequently

*crenati-, crenatus -a -um*   with small rounded teeth (the leaf margins, see Fig. 4(*a*))

*crepidatus -a -um*   sandal- or slipper-shaped

*Crepis*   a name usee by Theophrastus, meaning not clear

*crepitans*   rustling

*Crescentia*   for Pietro de Crescenzi of Bologna (1230–1321)

*cretaceus -a -um*   of chalk, inhabiting chalky soils

*creticus -a -um*   from Crete, Cretan

*criniger -era -erum*   carrying hairs

*Crinitaria*   Long-hair (the inflorescence)

*crinitus -a -um*   with a tuft of long soft hairs

*Crinodendron*   Lily-tree (floral similarity)

*crispatus -a -um*   closely waved, curled

*crispus -a -um*   with a waved or curled margin

*crista-galli*   cock's comb (the crested bracts)

*cristatus -a -um*   tassel-like at the tips, crested

*Crithmum*   Barley (the similarity of the seed)

*crocatus -a -um*   citron-yellow, saffron-like (used in dyeing)

*croceus -a -um*   saffron-coloured, yellow

*Crocosmia*   Saffron-scented (the dry flowers)

*Crocus*   from the Chaldean name

*Crotalaria*   Rattle (seeds loose in the inflated pods of some)

*Crucianella*   Little-cross

*Cruciata*   Cross (Dodoens' name refers to the arrangement of the leaves)

*cruciatus -a -um*   arranged cross-wise (leaf arrangement)

*crucifer -era -erum*   cross-bearing, cruciform

*cruentus -a -um*   blood-coloured, bloody

*cruentatus -a -um*   stained with red, bloodied

*crumenatus -a -um*   pouched

*crura*   legged

*crus, cruris*   leg, shin

*crus-andrae*   St Andrew's cross

*crus-galli*   cock's spur or leg

*crustatus -a -um*   encrusted

*crus-maltae, crux-maltae*   Maltese cross

*crypto-*   obscurely-, hidden-

*Cryptogramma(e)*   Hidden-writing (the concealed lines of sori)

*Cryptomeria*   Hidden-parts (the inconspicuous male cones)

*crystallinus -a -um*   with a glistening surface, as though covered with crystals

*cteno-*   comb-like-, comb-

*cubitalis -is -e*   a cubit tall (the length of the forearm plus the hand)

*Cucerbita*   the Latin name for the bottle-gourd

*Cucubalus*   a name in Dioscorides and Pliny

*cucallaris -is -e, cucullatus -a -um*   hood-like, hooded

*cucumerinus -a -um*   resembling cucumber, cucumber-like

*cucurbitinus -a -um*   melon- or marrow-like

*cujete*   a Brazilian name

*culinaris*   of food, of the kitchen

*cultoris, cultorus -a -um*   of gardeners, of gardens

*cultratus -a -um, cultriformis -is -e*   shaped like a knife-blade

*cultus -a -um*   cultivated, grown

*-culus -a -um*   -lesser

*-cundus -a -um*   -dependable, -able

*cuneatus -a -um, cuneiformis -is -e*   narrow below and wide above, wedge-shaped

*Cunninghamia*   for the discover of *C. lanceolata* in Chusan, China

*Cunonia*   for J. C. Cuno, Dutch naturalist (1708–1780)

*Cuphea*   Curve (the fruiting capsule's shape)

*cupreatus -a -um*   coppery, bronzed

*cupressinus -a -um, cupressoides*   cypress-like, resembling *Cupressus*

*Cupressus*   Symmetry (the conical shape). In mythology Apollo turned Kypressos into an evergreen tree

*cupreus -a -um*   copper-coloured, coppery

*cupularis -is -e*   cup-shaped

*curassavicus -a -um*   from Curaçao, West Indies

*curcas*   ancient Latin name for *Jatropha*

*Curculigo*   Weevil (the beak of the fruit)

*Curcuma*   the Arabic name for turmeric

*curti-, curtus -a -um*   broken-short, short

*curvatus -a -um, curvi-*   curved

*Cuscuta*   the medieval name for dodder

*cuspidatus -a -um, cuspidi-*   abruptly narrowed into a short rigid point (cusp), cuspidate

*cyaneus -a -um, cyano-, cyanus*   Prussian-blue

*Cyanotis*   Blue-ear

*Cycas*   Theophrastus' name for an unknown palm

*cycl-, cyclo-*   circle-, circular-

*Cyclamen*   Circled (the twisted fruiting stalk)

*cyclamineus -a -um*   resembling *Cyclamen*

*cyclius -a -um*   round, circular

*cyclops*   gigantic

*Cydonia*   the Latin name for an 'apple' tree from Cydon, Crete

*cylindricus -a -um, cylindro-*   long and round, cylindrical

*Cymbalaria*   Cymbal (the peltate leaves)

*cymbalarius -a -um*   cymbal-like (the leaves of toad flax)

*cymbi-, cymbidi-*   boat-shaped-, boat-

*Cymbidium*   Boat-like (the hollow recess in the lip)

*cymbiformis -is -e*   boat-shaped

*Cymbopogon*   Bearded-cup

*cymosus -a -um*   having flowers borne in a cyme (see Fig. 3(*a* to *d*)) *cynanchicus -a -um*   of quinsy (literally dog-throttling), from its former medicinal use

*cynanchoides*   resembling *Cynanchum*

*Cynanchum*   Dog-strangler (some are poisonous)

*Cynapium*   Dog-celery, Dog-parsley

*cyno-*   dog- (usually has derogatory undertone)

*cynobatifolius -a -um*   eglantine-leaved

*Cynodon*   Dog-tooth (the form of the spikelets)

*Cynoglossum*   Hound's-tongue

*cynops*   the ancient Greek name for a plantain

*Cynosurus*   Dog-tail

*cyparissias*   cypress-leaved (used in Pliny for a spurge)

*Cyperus*   the Greek name for several species

*Cypripedium*   Aphrodite's foot (Kypris was a name for Aphrodite or Venus)

*cyrt-*   curved-, arched-

*Cyrtomium*   Bulged (the leaflets)

*cyst-, cysti, cysto-*   hollow-, pouched-

*Cystopteris*   Bladder-fern

*Cytisus*   the Greek name for a clover-like plant


*Daboecia (Dabeocia)*   for St Dabeoc, Welsh missionery to Ireland

*dactyl-, dactylo-, dactyloides*   finger-, finger-like-

*Dactylis*   Grape-bunch (the inflorescence)

*Dactylorchis*   Finger orchid (the arrangement of the root-tubers)

*dahuricus -a -um, dauricus -a -um, davuricus -a -um*   from Dauria, north-east Asia, near Chinese border

*dalmaticus -a -um*   from Dalmatia, eastern Adriatic, Dalmatian

*damascenus -a -um*   from Damascus, coloured like *Rosa damascena*

*Damasonium*   a name in Pliny for *Alisma*

*Danae*   after the daughter of Acrisius Persius, in Greek mythology

*Danaea (Danaa)*   for J. P. M. Dana, Italian botanist (1734–1801)

*danfordiae*   for Mrs C. G. Danford

*danicus -a -um*   from Denmark, Danish

*Danthonia*   for Etienne Danthoine, student of the grasses of Provence, France

*Daphne*   old name for bay-laurel, from that of a nymph in Greek mythology

*daphnoides*   resembling *Daphne*

*dasy-*   thick-, thickly-hairy-

*dasyclados*   shaggy-twigged

*dasyphyllus -a -um*   thick-leaved
*-dasys*   -hairy
*Datura*   from an Indian vernacular name
*dauci-*   carrot-like, resembling *Daucus*
*Daucus*   the Latin name for a carrot
*Davidia, davidii, davidianus -a -um*   for l'Abbé Armand David,
   collector of Chinese plants (1826–1900)
*de-*   downwards-, outwards-, from-
*dealbatus -a -um*   with a white powdery covering, white-washed
*debilis -is -e*   weak, feeble, frail
*dec-, deca-*   ten-, tenfold-
*decalvans*   balding, becoming hairless
*decandrus -a -um*   ten-stamened
*deciduus -a -um*   not persisting beyond one season, deciduous
*dicipiens*   deceiving, deceptive
*declinatus -a -um*   turned aside, curved downwards
*decolorans*   staining, discolouring
*decompositus -a -um*   divided more than once (leaf structure)
*decoratus -a -um, decorus -a -um*   handsome, elegant, decorous
*decorticans, decorticus -a -um*   with shedding bark
*decumanus -a -um*   very large (refers literally to one tenth of a
   division of Roman soldiers)
*decumbens*   prostrate with tips turned up, decumbent
*decurrens*   running down, decurrent (e.g. the bases of leaves
   down the stem)
*decussatus -a -um*   at right-angles, decussate (as when the leaves
   are in two alternating ranks)
*deflexus -a -um*   bent sharply backwards
*deformis -is -e*   misshapen, deformed
*dehiscens*   splitting open, gaping, dehiscent
*dejectus -a -um*   debased
*delavayanus -a -um, delavayi*   for l'Abbé J. M. Delavay, collector
   of plants in China
*delectus -a -um*   choice, chosen
*delicatissimus -a -um*   most charming, most delicate
*deliciosus -a -um*   of pleasant flavour, delicious
*Delonix*   Conspicuous-claw (on the petals)
*delphicus -a -um*   from Delphi, Greece, Delphic
*Delphinium*   Dolphin-head (name used by Dioscorides)
*deltoides, deltoideus -a -um*   triangular-shaped, deltoid
*demersus -a -um*   underwater, submerged
*demissus -a -um*   hanging down, low, weak
*dendri-, dendro-*   tree-, tree-like-, on trees-

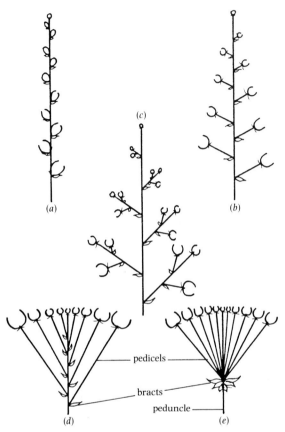

Fig. 2. Types of inflorescence which provide specific epithets:
(a) a spike (e.g. *Actaea spicata* L. and *Phyteuma spicatum* L.); (b) a
raceme (e.g. *Bromus racemosus* L. and *Sambucus racemosa* L.); (c) a
panicle (e.g. *Carex paniculata* L. and *Centaurea paniculata* L.); (d) a
corymb (e.g. *Silene corymbifera* Bertol. and *Teucrium corymbosum*
R.Br.); (e) an umbel (e.g. *Holosteum umbellatum* L. and *Butomus
umbellatus* L.).

In these inflorescences the oldest flowers are attached towards the
base and the youngest towards the apex.

*dendricolus -a -um*   tree-dwelling

*dendroideus -a -um, dendromorphus -a -um*   tree-like

*densatus -a -um, densi-, densus -a -um*   crowded, close, dense (habit of stem growth)

*dens-canis*   dog's tooth

*dens-leonis*   lion's tooth

*Dentaria*   Toothwort (the signature of the scales upon the roots)

*dentatus -a -um, dentifer -era -erum, dentosus -a -um*   having teeth, with outward-pointing teeth, dentate (see Fig. 4(*b*))

*denudatus -a -um*   hairy or downy but becoming naked, denuded

*deodarus -a -um*   from the Indian state of Deodar

*deorsus -a -um*   downwards, hanging

*deorum*   of the gods

*depauperatus -a -um*   imperfectly formed, dwarfed, of poor appearance, impoverished

*dependens*   hanging down, pendent

*depressus -a -um*   flattened downwards, depressed

*derelictus -a -um*   abandoned, neglected

*Deschampsia*   for the French naturalist M. H. Deschamps

*Descurania (Descurainia)*   for François Descourain, French physician (1658–1740)

*deserti-, desertorum*   of deserts

*desma-*   bundle-

*Desmanthus*   Bundle-flower (the appearance of the inflorescence)

*Desmazeria (Demazeria)*   for J. B. H. Desmazières, French botanist (1796–1862)

*detergens*   delaying

*detersus -a -um*   wiped clean

*detonsus -a -um*   shaved, bald

*deustus -a -um*   burned

*dextrorsus -a -um*   twining anticlockwise upwards as seen from outside

*di-, dis-*   two-, twice-, between-, away from-

*dia-*   through-, across-

*diabolicus -a -um*   slanderous, two-horned, devilish

*diacritus -a -um*   distinguished, separated

*diadema, diadematus -a -um*   band or fillet, crown, crown-like

*dialy-*   very deeply incised-, separated-

*diandrus -a -um*   two-stamened

*Dianthus*   Jove's flower (a name used by Theophrastus)

*Diapensia*   formerly a name for sanicle but reapplied by Linnaeus

*diaphanoides*   resembling *Hieracium diaphanum* (in leaf form)

93

*diaphanus -a -um*   transparent (leaves)

*Dicentra*   Twice-spurred (the two-spurred flowers)

*dicha-, dicho-*   double-, into two-

*Dichondra*   Two-lumped (the two-lobed ovary)

*Dichorisandra*   Two-separated-men (two of the stamens diverge from the remainder)

*dichotomus -a -um*   repeatedly divided into two equal portions

*dichrano-*   two-branched-

*dichranotrichus -a -um*   with two-pointed hairs

*dichroanthus -a -um*   with two-coloured flowers

*dichromatus -a -um, dichromus -a -um, dichrous -a -um*   of two colours, two coloured

*dicoccus -a -um*   having paired nuts, two-berried

*dictyo-*   netted-

*didymo-, didymus -a -um*   twin-, twinned-, double, equally-divided, in pairs

*dielsianus -a -um, dielsii*   for F. L. E. Diels of the Berlin Botanic Garden (1874–1945)

*Dierama*   Funnel (the shape of the perianth)

*difformis -is -e*   of unusual form or shape

*diffusus -a -um*   loosely spreading, diffuse

*Digitalis*   Thimble (from the German 'Fingerhut')

*Digitaria*   Fingered (the radiating spikes)

*digitatus -a -um*   fingered, hand-like, lobed from one point, digitate

*Digraphis, digraphis -is -e*   twice inscribed, with lines of two colours

*dilatatus -a -um, dilatus -a -um*   widened, spread out, dilated

*dilectus -a -um*   precious, valuable

*dilutus -a -um*   washed

*dimidiatus -a -um*   with two equal parts, dimidiate

*dimorpho-, dimorphus -a -um*   two-shaped, with two forms (of leaf or flower or fruit)

*Dimorphotheca*   Two-kinds-of-container (the fruits vary in shape)

*dinaricus -a -um*   from the Dinaric Alps

*diodon*   two-toothed

*dioicus -a -um*   of two houses, having separate male and female plants, dioecious

*Dioscorea*   for Dioscorides (type genus of the yam family)

*Diosma*   Divine-fragrance

*Diospyros*   Divine-fruit

*Diotis*   Two-ears (the spurs of the corolla)

*Dipcadi*   from an oriental name for *Muscari*

*diphyllus -a -um*   two-leaved
*Diplachne*   Double-chaff
*Diplazium*   Duplicate (the double indusium)
*Diplopappus*   Double-down
*Diplotaxis*   Two-positions (the two-ranked seeds)
*diplotrichus -a -um*, *diplothrix*   having two kinds of hairs
*dipsaceus -a -um*   teasel-like, resembling *Dipsacus*
*Dipsacus*   Dropsy (analogy of the water-collecting leaf-bases)
*diptero-*, *dipterus -a -um*   two-winged
*Disanthus*   Paired-flowers (they are in two's)
*discerptus -a -um*   disc-like, discoid
*disci- disco-*   disk-
*discipes*   with a disc-like stalk
*discolor*   of different colours
*disjunctus -a -um*   separated, not grown together, disjunct
*dispar*   unequal, different
*dispersus -a -um*   scattered
*dissectus -a -um*   cut into many deep lobes
*dissimilis -is -e*   unlike
*dissitiflorus -a -um*   with flowers not in compact heads
*distachyon*, *distachyus -a -um*   two-branched, two-spiked, with
   two spikes
*distans*   widely separated, distant
*distichus -a -um*   in two opposed ranks (leaves or flowers)
*distillatorius -a -um*   shedding drops, of the distillers
*distortus -a -um*   malformed, grotesque, distorted
*Distylium*   Two-styles (the separate styles)
*distylus -a -um*   two-styled
*diurnus -a -um*   lasting for one day, day-flowering, of the day
*diutinus -a -um*   long-lasting
*divaricatus -a -um*   wide-spreading, straggling, divaricate
*divensis -is -e*   from Chester (Deva)
*divergens*   spreading out, wide-spreading, divergent
*diversi-*   differing-, variable-, diversely-
*divionensis -is -e*   from Dijon, France
*divisus -a -um*   divided
*divulsus -a -um*   torn violently apart
*divus -a -um*   belonging to the gods
*dodec-*   twelve-
*Dodecatheon*   Twelve-gods (an ancient name)
*Dodonaea*   for Rembert Dodoens, botanical writer (1518–1585)
*dolabratus -a -um*, *dolabriformis -is -e*   hatchet-shaped
*dolich-*   long-

*Dolichos*   the ancient Greek name for long-podded beans
*dolichostachyus -a -um*   long-spiked
*dolobratus -a -um*   hatchet-shaped, see *dolabratus*
*dolosus -a -um*   deceitful
*domesticus -a -um*   of the household
*donax*   an old Greek name for a reed
*Doronicum*   from an Arabic name
*Dorotheanthus*   Dorothea-flower (for Dr Schwantes' mother, Dorothea)
*-dorus -a -um*   -bag-shaped, -bag
*dory-*   spear-
*Dorycnium*   ancient Greek name for a *Convolvulus* reapplied by Dioscorides
*Doryopteris*   Spear-fern
*Douglasia*   for David Douglas, collector in American North-West for the Royal Horticultural Society (1798–1834)
*-doxa*   -glory
*Draba*   a name used by Dioscorides for *Lepidium draba*
*drabi-*   *Draba*-like
*Dracaena*   Female-dragon
*Dracunculus*   Little-dragon (a name used by Pliny)
*drepano-*   sickle-shaped
*Drepanocarpus*   Curved-fruit (Leopard's claw)
*drepanus -a -um*   from a town in west Sicily
*Drimia*   Acrid (the pungent juice from the roots)
*Drimys*   Acrid (the taste of the bark)
*Drosanthemum*   Dewy-flower (glistens with epidermal hairs)
*Drosera*   Dew (the glistening glandular hairs)
*drucei*   for George Claridge Druce, British botanist (1859–1932)
*drupaceus -a -um*   stone-fruited, drupe-like
*Dryas*   Oak-nymph (the leaf shape)
*drymo-*   wood-, woody-
*dryophyllus -a -um*   oak-leaved
*Dryopteris*   Oak-nymph-fern (the woodland habitat)
*dubius -a -um*   uncertain, doubtful
*Duchesnea*   for A. N. Duchesne, French botanist (1747–1827)
*dulcamara*   bitter-sweet
*dulcis -is -e*   sweet-tasted, mild
*dumalis -is -e, dumosus -a -um*   compact, thorny, bushy
*dumetorum*   of bushy habitats, of thickets
*dumnoniensis -is -e*   from Devon, Devonian
*dunensis -is -e*   of sand-dunes
*duplex, duplicatus -a -um*   growing in pairs, double, duplicate

*duracinus -a -um*   hard-fruited, hard-berried

*Durio*   from the Malaysian name for the fruit

*durior, durius*   harder

*duriusculus -a -um*   rather hard or rough

*durmitoreus -a -um*   from the Durmitor mountains, Yugoslavia

*durus -a -um*   hard

*Durvillaea*   for J. S. C. D. d'Urville, French naval officer (1790–1842)

*Dyckia*   for Prince Salm-Dyck, German writer on succulent plants (1773–1861)

*dys-*   poor-, ill-, bad-, difficult-

*Dyschoriste*   Poorly-divided (the stigma)

*dysentericus -a -um*   of dysentery (medicinal treatment for)

*dyso-*   evil-smelling-

*Dysodea*   Evil-scented


*e-, ef-, ex-*   without-, not-, from out of-

*ebenaceus -a -um*   ebony-like

*ebenus -a -um*   ebony-black

*eboracensis -is -e*   from York (Eboracum)

*ebracteatus -a -um*   without bracts

*Ebulus*   a name in Pliny for danewort

*eburneus -a -um*   ivory-white

*ecae*   for Mrs E. C. Aitchison

*Ecballium*   Expeller (the discharging fruit of the squirting cucumber)

*Eccremocarpus*   Hanging-fruit

*Echeveria*   For Atanasio Echeverria, botanical artist

*echinatus -a -um, echino-*   covered with prickles, hedgehog-like

*Echinochloa*   Hedgehog-grass (the awns)

*Echinodorus*   Hedgehog-bag (the fruiting heads of some species)

*Echinops*   Hedgehog-resembler

*echioides*   resembling *Echium*

*Echium*   Viper (a name used by Dioscorides)

*electus -a -um*   picked out, selected

*ect-, ecto-*   on the outside-, outwards-

*edentatus -a -um*   without teeth, toothless

*editorum*   of the editors

*Edraianthus*   Sessile-flower

*edulis -is -e*   of food, edible

*effusus -a -um*   spread out, very loose-spreading

*Eglanteria, eglantarius -a -um*   from a French name (eglantois or eglanties)

*Eichornia (Eichhornia)*   for J. A. F. Eichhorn of Prussia
   (1779–1856)
*Elaeagnus*   Olive-chaste-tree
*Elaeis*   Oil (copious in the oil-palm fruit)
*elaeo-*   olive-
*elapho-*   stag's-
*Elaphoglossum*   Stag's-tongue (shape and texture of the fronds)
*elasticus -a -um*   yielding an elastic substance, elastic
*elatarius -a -um*   driving away (squirting out seeds)
*Elatine*   Little-fir-trees (a name used by Dioscorides)
*elatior, elatius*   taller
*elatus -a -um*   tall, high
*elegans, elegantulus -a -um*   graceful, elegant
*eleo-*   marsh (cf. *heleo-*)
*Eleocharis (Heleocharis)*   Marsh-favour
*Eleogiton (Heleogiton)*   Marsh-neighbour (in analogy with
   *Potamogeton*)
*elephantidens*   elephant's tooth
*elephantipes*   like an elephant's foot (the appearance of the
   stem)
*Eleusine*   from Eleusis, Greece
*eleuthero-*   free-
*Elisma*   a variant of *Alisma*
*Elliottia*   for Stephen Elliott, American botanist (1771–1830)
*elliottii*   for either G. M. Scott-Elliott, botanist in Sierra Leone
   and Madagascar, or Capt. Elliott, plant grower of Farnborough
   Park, Hants
*ellipsoidalis -is -e*   ellipsoidal (a solid of oval profile)
*ellipticus -a -um*   about twice as long as broad, elliptic
*-ellus -ella -ellum*   -lesser (diminutive ending)
*Elodea*   Marsh (growing in water)
*elodes*   as *helodes*, of bogs and marshes
*elongatus -a -um*   lengthened out, elongated
*Elymus*   Hippocrates' name for a millet-like grass
*em-, en-*   in-, into- within-, for-, not-
*emarginatus -a -um*   notched at the apex (see Fig. 7(*h*))
*emasculus -a -um*   without functional stamens
*Embothrium*   In-little-pits (position of its anthers)
*emerus*   from an early Italian name for a vetch
*emeticus -a -um*   causing vomiting, emetic
*eminens*   noteworthy, outstanding, prominent
*emodensis -is -e, emodi*   from the western Himalayas, 'Mt
   Emodus', north India

*Empetrum*   In-rocks (Dioscorides' name refers to the habitat)
*Enarthocarpus*   Jointed-fruit
*encephalo-*   in a head-
*Encephalartos*   In-a-head-bread (farinaceous centre of the stem yields sago, as in sago-palms)
*endivia*   ancient Latin name for chicory (see *Intybus*)
*Endymion*   for Diana's lover, of Greek mythology
*enervis -is -e*   apparently lacking nerves (veins)
*Englera, Englerastrum, Englerella*   for Adolf Engler, director of Dahlem Botanic Garden (1844–1930)
*enki-*   swollen-
*Enkianthus*   Swollen-flower (the corolla of some)
*ennea-*   nine-
*enneagonus -a -um*   nine-angled
*ensatus -a -um, ensi-*   sword-shaped (leaves)
*-ensis -is -e*   -from, -of (after the name of a place)
*ento-, endo-*   on the inside-, inwards-
*entomophilus -a -um*   of insects, insect-loving
*ep-, epi-*   upon-, on-, over-, somewhat-
*Epacris*   Upon-the-summit (some live on hilltops)
*Ephedra*   from a name in Pliny for *Hippuris*
*epiteius -a -um*   annual
*Epidendron(um)*   Tree-dweller (the epiphytic habit)
*epigaeus -a -um*   growing close to the ground's surface
*epigeios*   of dry earth, from dry habitats
*epihydrus -a -um*   of the water surface
*Epilobium*   Gesner's name indicating the positioning of the corolla on top of the ovary
*Epimedium*   the name used by Dioscorides
*Epipactis*   a name used by Theophrastus
*epiphyllus -a -um*   upon the leaf (flowers or buds)
*epiphyticus -a -um*   growing upon another plant
*Epipogium (Epipogon)*   Over-beard (the lip of the ghost-orchid is uppermost)
*epipsilus -a -um*   somewhat naked (the sparse foliage)
*epistomius -a -um*   snouted (flowers)
*epithymoides*   thyme-like
*epithymum*   upon thyme (parasitic)
*equalis -is -e*   equal
*equestris -is -e*   of horses or horsemen, equestrian
*equinus*   of the horse
*Equisetum*   Horse-hair (a name in Pliny for a horsetail)
*equitans*   astride as on horseback (leaves of *Iris*)

*Eragrostis*   Love-grass

*Eranthemum*   Beautiful-flower

*Eranthis*   Spring-flower (early flowering season)

*erectus -a -um*   upright, erect

*eremi-, eremo-*   deserted-, solitary-

*Eremurus*   Solitary-tail (the long raceme)

*eri-, erio-*   woolly-

*Erica*   Pliny's version of a name used by Theophrastus

*ericetorum*   of heathland

*ericinus -a -um, ericoides, erici-*   heath-like, resembling *Erica*

*erigenus -a -um*   Irish born

*Erigeron*   Early-old-man (Theophrastus' name)

*erinaceus -a -um*   hedgehog-like, prickly

*Erinus*   Dioscorides' name for an early flowering basil-like plant

*Eriocaulon*   Woolly-stem

*Eriophorum*   Wool-bearer (cotton grass)

*eriophorus -a -um*   bearing wool

*eristhales*   very luxuriant, *Eristhalis*-like

*ermineus -a -um*   ermine-coloured

*Erodium*   Heron (the shape of the fruit)

*Erophila*   Spring-lover

*erosus -a -um*   jagged, as if nibbled irregularly, erose

*erraticus -a -um*   differing from the type, of no fixed habitat

*erromenus -a -um*   vigorous, strong, robust

*erubescens*   blushing, turning red

*Eruca*   Belch (the ancient Latin name for cole-wort)

*Erucastrum*   *Eruca*-flowered

*Ervum*   the Latin name for *Orobus*, a vetch

*Eryngium*   Theophrastus' name for a spiny-leaved plant

*Erysimum*   a name used by Theophrastus

*Erythea*   for the daughter of night and the dragon Lado of
   mythology, one of the Hesperides

*erythraeus -a -um, erythro-*   red

*Erythronium*   Red (flower colours)

*Erythroxylon*   Red-wood

*Escallonia*   for the Spanish South American traveller named
   Escallon

*Eschscholzia (Eschscholtzia)*   for J. F. Eschscholtz, traveller and
   naturalist (1793–1831)

*-escens*   -becoming, -ish

*esculentus -a -um*   tasty, good to eat, edible, esculent

*estriatus -a -um*   without stripes

*esula*   an old generic name from Rufinus

*-etorus -a -um*   -community (indicating the habitat)

*etruscus -a -um*   from Tuscany (Etruria), Italy

*ettae*   for Miss Etta Stainbank

*eu-*   good-, proper-, completely-, well-marked

*Euadenia*   Well-marked-glands (the five lobes at the base of the gynophore)

*euboeus -a -um, euboicus -a -um*   from the Greek island of Euboea

*Eucalyptus*   Complete-cover (the calyx)

*Eucharis*   Full-of-grace

*euchlorus -a -um*   of beautiful green, true green

*euchromus -a -um*   well-coloured

*Euclidium*   Well-closed (the fruit)

*Eucomis*   Beautiful-head

*eudoxus -a -um*   of good character

*Eugenia*   for Prince Eugene of Savoy, patron of botany (1663–1736)

*eugenioides*   *Eugenia*-like

*Eulophia*   Beautiful-crest (the crests of the lip)

*euodes*   sweet-scented

*Euonymus*   Famed (Theophrastus' name, also appears as *Evonymus*)

*Eupatorium*   for Eupator who was King of Pontus

*euphlebius -a -um*   well-veined

*Euphorbia*   for Euphorbus who used the latex for medicinal purposes

*Euphrasia*   Good-cheer (signature of eyebright flowers as being of use in eye lotions)

*euphues*   well-grown

*eupodus -a -um*   long-stalked

*euprepes, eupristis -a -um*   comely, good-looking

*Eupteris*   Proper-*Pteris*

*eur-, euro-, eury-*   wide-, broad-

*europaeus -a -um*   from Europe, European

*Euryale*   from one of the Gorgons of mythology (had burning thorns in place of hair)

*-eus -ea -eum*   -resembling, -belonging to, -noted for

*eustachyus -a -um*   having long trusses of flowers

*Euterpe*   Attractive (the name of one of the muses)

*evanescens*   quickly disappearing, evanescent

*evectus -a -um*   lifted up, springing out

*evertus -a -um*   overturned, expelled, turned out

*Evodia (Euodia)*   Well-perfumed

*ex-*   without-, outside-, over and above-

*Exaculum*   *Exacum*-like

*Exacum*   a name in Pliny (may be derived from an earlier Gallic word, or refer to its expulsive property)

*exalbescens*   out of *albescens* (related to)

*exaltatus -a -um*   lofty, very tall

*exaratus -a -um*   with embossed grooves, engraved

*exasperatus -a -um*   rough, roughened (surface texture)

*excavatus -a -um*   hollowed out, excavated

*excellens*   distinguished, excellent

*excelsior, excelsus -a -um*   higher, taller, very tall

*excisus -a -um*   cut away, cut out

*excorticatus -a -um*   without bark, stripped (without cortex)

*excurrens*   with a vein extended into a marginal tooth (as on some leaves)

*exiguus -a -um*   very small, meagre, poor, petty

*exili-, exilis -is -e*   meagre, small, few, slender, thin

*eximius -a -um*   excellent in size or beauty, choice, distinguished

*exitiosus -a -um*   fatal, deadly, pernicious, destructive

*Exochorda*   Outside-cord (the vascular anatomy of the wall of the ovary)

*exoniensis -is -e*   from Exeter, Devon

*exotericus -a -um*   common, external

*exoticus -a -um*   from a foreign land, not native, exotic

*expansus -a -um*   spread out, expanded

*expatriatus -a -um*   without a country

*explodens*   exploding

*exscapus -a -um*   with a stem

*exsculptus -a -um*   with deep cavities, dug out

*exsectus -a -um*   cut out

*exsertus -a -um*   projecting, protruding, held out

*extensus -a -um*   wide, extended

*extra-*   outside-, beyond-, over and above-

*extrorsus -a -um*   directed outwards from the central axis (outwards facing stamens) extrorse

*exudans*   producing a (sticky) secretion, exuding

*faba*   the old Latin name for the broad bean

*fabaceus -a -um, fabarius -a -um*   bean-like, resembling *Faba*

*facetus -a -um*   elegant, fine, humorous

*faenum*   hay, fodder

*fagi-*   beech-like, *Fagus-*

*Fagopyrum(on)*   Beech-wheat (buckwheat is from the Dutch Boekweit)

*Fagus*   the Latin name for the beech tree

*Falcaria*   Sickle (the shape of the leaf-segments)

*falcatus -a -um, falci-, falciformis -is -e*   sickle-shaped, falcate

*fallax*   deceitful, deceptive, false

*Faradaya*   for Michael Faraday, scientist (1794–1867)

*farcatus -a -um*   solid, not hollow

*farfara*   an old generic name for butterbur

*fargesii*   for Paul Guillaume Farges, collector in Szechwan, China

*farinaceus -a -um*   of mealy texture, yielding farina (starch), farinaceous

*farinosus -a -um*   with a mealy surface, mealy, powdery

*farleyensis -is -e*   from Farley Hill Gardens, Barbados, West Indies

*farnesianus -a -um*   from the Farnese Palace gardens of Rome

*farreri*   for Reginald J. Farrer, English plant hunter (1880–1920)

*fasciarus -a -um*   elongate and with parallel edges, band-shaped

*fasciatus -a -um*   bound together, bundled, fasciated as in the inflorescence of cockscomb *Celosia*

*fascicularis -is -e, fasciculatus -a -um*   clustered in bundles, fascicled

*fastigiatus -a -um*   with erect branches, fastigiate

*fastuosus -a -um*   proud

*Fatsia*   from a Japanese name

*fatuus -a -um*   not good, insipid, simple, foolish

*faucilalis -is -e*   wide-mouthed

*favigerus -a -um*   bearing honey-glands

*favosus -a -um*   cavitied, honeycombed

*febrifugus -a -um*   fever-dispelling (medicinal property)

*fecundus -a -um*   fruitful, fecund

*fejeensis -is -e*   from the Fiji Islands

*Felicia*   for a German official named Felix, some interpret it as Cheerful

*felix*   fruitful

*felosmus -a -um*   foul-smelling

*femina*   feminine

*fenestralis -is -e, fenestratus -a -um*   with window-like holes or openings

*fennicus -a -um*   from Finland (Fennica), Finnish

*-fer, -ferus -a -um*   -bearing, -carrying

*ferax*   fruitful

*fero-, ferus -a -um*   wild, feral

*ferox*   very prickly, ferocious
*ferreus -a -um*   durable, iron-hard
*ferrugineus -a -um*   rusty-brown in colour
*ferruginosus -a -um*   conspicuously brown
*fertilis -is -e*   heavy-seeding, fruitful, fertile
*Ferula*   Rod (the classical Latin name)
*ferulaceus -a -um*   fennel-like, resembling *Ferula*
*festalis -is -e, festinus -a -um, festivus -a -um*   agreeable, bright,
   pleasant, cheerful, festive
*Festuca*   Straw (a name used in Pliny)
*festus -a -um*   sacred, used for festivals
*fetidus -a -um*   bad smelling, stinking, foetid
*fibrillosus -a -um, fibrosus -a -um*   with copious fibres, fibrous
*ficaria*   small-fig, and old generic name for the lesser celandine
   (the root tubers)
*fici-, ficoides*   fig-like, resembling *Ficus*
*ficto-, fictus -a -um*   false
*Ficus*   the ancient Latin name for the fig
*-fid, -fidus -a -um*   -cleft, -divided
*Filago*   Thread (the medieval name refers to the woolly
   indumentum)
*filamentosus -a -um, filarius -a -um, fili-*   thread-like, with
   filaments or threads
*filicaulis -is -e*   having very slender stems
*filicinus -a -um, filici-, filicoides*   fern-like
*filiculoides*   like a small fern
*filiferus -a -um*   bearing threads or filaments
*filiformis -is -e*   thread-like
*Filipendula*   Thread-suspended (slender attachment of the
   tubers)
*filix*   fern
*filix-femina (foemina)*   female fern
*filix-mas*   male fern
*fimbriatus -a -um, fimbri-*   fringed
*firmus -a -um*   strong, firm
*fissi-, fissilis -is -e, fissuratus -a -um, fissus -a -um*   cleft, divided
*fissus -a -um*   cleft almost to the base
*fistulosus -a -um*   hollow, pipe-like, tubular, fistular
*Fittonia*   for E. and S. M. Fitton, botanical writers
*flabellatus -a -um*   fan-like, fan-shaped, flabellate
*flabellifer -era -erum*   fan-bearing (with flabellate leaves)
*flabelliformis -is -e*   pleated fanwise
*flaccidus -a -um*   limp, weak, feeble, soft, flaccid

*flaccus -a -um*   drooping, flabby

*Flacourtia*   for Etienne de Flacourt, French East India Company (1607–1661)

*flagellaris -is -e, flagellatus -a -um, flagelli-*   with long thin shoots, whip-like, stoloniferous

*flagelliferus -a -um*   bearing whips (elongate stems of New Zealand trip-me-up sedge)

*flammeus -a -um*   flame-red, fiery-red

*flammula*   an old generic name for lesser spearwort

*flammulus -a -um*   flame-coloured

*flavens*   being yellow

*flav-, flavi-, flaveolus -a -um, flavo-*   yellowish

*flavescens*   pale-yellow, turning yellow

*flavidus -a -um*   yellowish

*flavus -a -um*   bright almost pure yellow

*flexi-, flexilis -is -e*   pliant, flexible

*flexuosus -a -um*   zig-zag, winding, much bent, tortuous

*-flexus -a -um*   -turned

*flocciger -era, -erum, floccosus -a -um*   with a woolly indumentum which falls away in tufts, floccose

*flora*   the Roman goddess of flowering plants

*flore-albo*   white-flowered

*florentinus -a -um*   from Florence, Florentine

*flore-pleno*   double-flowered, full-flowered

*floribundus -a -um*   abounding in flowers, freely-flowering

*floridanus -a -um*   from Florida, USA

*floridus -a -um*   free-flowering, flowery

*florindae*   for Mrs Florinda N. Thompson

*florulentus -a -um*   flowery

*-florus -a, -um*   -flowered

*flos-cuculi*   cuckoo-flowered, flowering in the season of cuckoo song

*flos-jovis*   Jove's flower

*fluctuans*   inconstant, fluctuating

*fluitans*   floating on water

*fluminensis -is -e*   growing in running water, of the river

*fluvialis -is -e, fluviatilis -is -e*   growing in rivers and streams

*foecundus -a -um*   fruitful, fecund

*foemina*   feminine

*Foeniculum*   the Latin name for fennel

*foenisicii*   of mown hay

*foetidus -a -um, foetens*   stinking, bad smelling, foetid

*foliaceus -a -um*   leaf-like

*foliatus -a -um, foliosus -a -um*   leafy

*-foliatus -a -um*   -leaflets

*-folius -a -um*   -leaved

*follicularis -is -e*   bearing follicles (seed capsules as in hellebores)

*fontanus -a -um, fontinalis -is -e*   of fountains springs or fast-running streams

*forficatus -a -um*   scissor-shaped, shear-shaped (leaves)

*formicarius -a -um*   relating to ants

*-formis -is -e*   -resembling, -shaped

*formosanus -a -um*   from Taiwan (Formosa)

*formosus -a -um*   handsome, beautiful, well-formed

*fornicatus -a -um*   arched

*forrestii*   for George Forrest, plant hunter in China (1873–1932)

*forsteri, forsterianus -a -um*   for J. R. Forster or his son, J. G. A. Forster, of Halle

*Forsythia*   for William Forsyth of Kensington Royal Gardens (1737–1804)

*fortis -is -e*   strong

*fortunatus -a -um*   rich, favourite

*fortunei*   for Robert Fortune, collector for the Royal Horticultural Society in China (1812–1880)

*foulaensis -is -e*   from Foula, Scotland

*foveolatus -a -um*   with small depressions or pits all over the surface, foveolate

*fragari-, fragi-*   strawberry-

*Fragaria*   Fragrance (the fruit)

*fragifer -era -erum*   strawberry-bearing

*fragilis -is -e*   brittle, fragile

*fragrans*   sweet-scented, odorous, fragrant

*frainetto*   from a local name for an oak in the Balkans

*franciscanus -a -um, fransiscanus -a -um*   from San Francisco, USA

*Frangula*   Fragile (medieval name refers to the brittle twigs of alder buckthorn)

*Frankenia*   for John Frankenius, Swedish botanist (1590–1661)

*fraternus -a -um*   closely related, brotherly

*fraxineus -a -um*   ash-like, resembling ash

*Fraxinus*   ancient Latin name used by Virgil

*Freesia*   for F. H. T. Freese, pupil of Ecklon

*frene-*   strap-

*fresnoensis -is -e*   from Fresno County, California, USA
*frigidus -a -um*   cold, of cold habitats, of cold regions
*friscus -a -um*   Friesian
*Fritillaria*   Dice-box (the shape of the flowers)
*frondosus -a -um*   leafy
*fructifer -era -erum*   fruit-bearing, fruitful
*fructu-*   fruit-
*frumentaceus -a -um*   grain-producing
*frutescens, fruticans, fruticosus -a -um*   shrubby, becoming
   shrubby
*frutex*   shrub, bush
*fruticulosus -a -um*   dwarf-shrubby
*fucatus -a -um*   painted, dyed
*Fuchsia*   for Leonard Fuchs, German renaissance botanist
   (1501–1566)
*fucifer -era -erum*   drone-bearing
*fuciformis -is -e, fucoides*   bladder-wrack-like, resembling *Fucus*
   (seaweed)
*fugax*   fleeting, rapidly withering, fugacious
*Fuirena*   for G. Fuiren, Danish physician
*fulgens, fulgidus -a -um*   shining, glistening (often with red
   flowers)
*fuliginosus -a -um*   dirty-brown to blackish, sooty
*fullonum*   of cloth fullers
*fulvescens*   becoming tawny
*fulvi-, fulvus -a -um*   tawny, reddish-yellow, fulvous
*Fumaria*   Smoke-of-the-earth (medieval name refers to the smell
   of some and the hazy colour effect)
*fumidus -a -um*   smoke-coloured, dull grey coloured
*fumosus -a -um*   smoky
*funalis -is -e*   twisted together, rope-like
*funebris -is -e*   mournful, doleful, of graveyards, funereal
*fungosus -a -um*   spongy, fungus-like, pertaining to fungi
*funiculatus -a -um*   like a thin cord
*furcans, furcatus -a -um*   forked, furcate
*Furcraea*   for A. T. Fourcroy, French chemist (1755–1809)
*furfuraceus -a -um*   scurfy, mealy, scaly
*furiens*   exciting to madness
*fuscatus -a -um*   somewhat dusky-brown
*fusci-, fusco-, fuscus -a -um*   bright-brown, swarthy, dark-
   coloured
*fusiformis -is -e*   spindle-shaped
*futilis -is -e*   useless

*gaditanus -a -um* from Cadiz, Spain

*Gagea* for Sir Thos. Gage, English botanist (1781–1820)

*Gaillardia* for Gaillard de Charentonneau (Marentonneau), patron of botany

*galactinus -a -um* milky

*Galanthus* Milk-flower (the colour)

*galbanifluus -a -um* with a yellowish exudate (some *Ferula* species yield gum galbanum)

*galbinus -a -um* greenish-yellow

*gale* from an old English vernacular name for bog-myrtle or sweet gale

*galeatus -a -um, galericulatus -a -um* helmet-shaped, like a skull-cap

*Galega* Milk-promoter

*galegi-* resembling *Galega*

*Galeobdolon* Weasel-smell (a name used in Pliny)

*Galeopsis* Weasel-like (an ancient Greek name)

*Galinsoga* for Don M. Martinez de Galinsoga, Spanish botanist

*galioides* bedstraw-like, resembling *Galium*

*Galium* Milk (the flowers of *G. verum* were used to curdle milk in cheesemaking)

*gallicus -a -um* from France, French, of the cock or rooster

*Galphimia* anagram of Malpighia

*gamo-* fused-, united-, married-

*gandavensis -is -e* from Ghen, Belgium

*gangeticus -a -um* from the Ganges region

*Gardenia* for Dr A. Garden, American correspondent with Linnaeus (1730–1791)

*gardneri, gardnerianus -a -um* for Hon. E. Gardner (Nepal) or G. Gardner (Brazil)

*garganicus -a -um* from Mount Gargano, southern Italy

*Garrya* for Nicholas Garry of the Hudson's Bay Company

*gaster-, gastro-* belly-, bellied-

*Gasteria* Belly (the swollen base on the corolla)

*Gastridium* Little-paunch (the bulging of the glumes)

*Gaudinia* for J. F. G. P. Gaudin, Swiss botanist (1766–1833)

*Gaultheria* for Dr Gaulthier, Canadian botanist

*Gaura* Superb

*gayanus -a -um* for Jacques É. Gay, French botanist (1786–1864)

*Gaylussacia* for J. L. Gay-Lussac, French chemist (1778-1850)

*Gazania* for Theodore of Gaza who transcribed Theophrastus'

works into Latin (1398–1478), some interpret it as Riches
  (Lat. *gaza -ae*)
*geito-, geitono-* neighbour-
*gelidus -a -um* of icy regions, growing in icy places
*gemellus -a -um* in pairs, paired, twinned
*geminatus -a -um, gemini-* united in pairs, twinned
*gemmatus -a -um* jewelled
*gemmiferus -a -um, gemmiparus -a -um* bearing buds or
  propagules
*genavensis -is -e, genevensis -is -e* from Geneva, Switzerland
*generalis -is -e* normal, prevailing, usual
*geniculatus -a -um* with a knee-like bend
*Genista* a name in Virgil (*planta genista* from which the
  Plantagenets took their name)
*genisti-* broom-like, resembling *Genista*
*Gentiana* a name in Pliny (for Gentius, a King of Illyria)
*Gentianella* Gentian-like
*gentilis -is -e* foreign, of the same race, noble
*genuinus -a -um* natural, true
*geo-* on or under the earth-
*geocarpus -a -um* with fruits which ripen underground
*geoides* *Geum*-like
*geometrizans* equal, symmetrical
*geophilus -a -um* spreading horizontally, ground-loving
*georgei* for George Forrest, collector in China (1873–1932)
*georgianus -a -um* from Georgia, USA
*georgicus -a -um* from Georgia, Caucasus, USSR
*Geranium* Crane (Dioscorides' name refers to the shape of the
  fruit resembling the head of a crane)
*Gerardia* for John Gerard, author of the Herbal of 1597
  (1545–1612)
*Gerbera* for Traugott Gerber, German traveller
*germanicus -a -um* from Germany, German
*Gesneria* for Conrad Gesner, German botanist (1516–1565)
*-geton* -neighbour
*Geum* a classical name in Pliny
*gibb-, gibbi, gibbosus -a -um* swollen or enlarged on one side,
  gibbous
*gibberosus -a -um* humped, hunchbacked
*gibraltaricus -a -um* from Gibralter
*giganteus -a -um* unusually large or tall, gigantic
*giganthes* giant-flowered

*gigas*   giant

*gileadensis -is -e*   from Gilead, an Egyptian mountain range

*Gilia*   from a Hottentot name for a plant used to make a
   beverage

*giluus -a -um, gilvo-, gilvus -a -um*   dull pale yellow

*gingidium*   from an old name used by Dioscorides

*Ginkgo*   derived from a Japanese name (*gin-kyo*)

*githago*   an old generic name (green with red-purple stripes)

*glabellus -a -um*   somewhat smooth, smoothish

*glaber -ra -rum, glabro*   smooth, without hairs, glabrous

*glaberrimus -a -um*   very smooth, smoothest

*glabratus -a -um, glabrescens*   becoming smooth or glabrous

*glabriusculus -a -um*   rather glabrous

*glacialis -is -e*   of the ice, of frozen habitats

*gladiatus -a -um*   sword-like

*Gladiolus*   Small-sword (the leaves)

*glandulifer -era -erum*   gland-bearing

*glandulosus -a -um*   full of glands, glandular

*glaucescens, glaucus -a -um*   with a fine whitish bloom, bluish-
   green, sea-green, glaucous

*glaucifolius -a -um*   with grey-green leaves

*glauciifolius -a -um*   with leaves resembling those of horned
   poppy, *Glaucium*

*Glaucium*   Grey-green

*glaucophyllus -a -um*   glaucous-leaved

*glaucopsis -is -e*   glaucous-looking

*Glaux*   a name used by Dioscorides

*Glechoma*   Dioscorides' name for penny-royal

*Gleditsia (Gleditschia)*   for J. G. Gleditsch (Gleditsius) of the Berlin
   Botanic Garden (1714–1786)

*Gleichenia*   for F. W. Gleichen, German botanist

*Gliricidia*   Mouse-killer (the poisonous seed and bark)

*glischrus -a -um*   sticky, gluey, glandular bristly

*globatus -a -um*   arranged or collected into a ball

*globosus -a -um, globularis -is -e*   with small spherical parts,
   spherical (e.g. flowers)

*globulifer -era -erum*   carrying small balls (the sporocarps of
   pillwort)

*globulosus -a -um*   small round-headed

*glochidiatus -a -um*   with short barbed detachable bristles

*glomeratus -a -um*   collected into heads, aggregated, glomerate

*glomerulatus -a -um*   with small clusters or heads

*gloriosus -a -um*   superb, full of glory

*glosso-, glossus -a -um, -glottis*   tongue-shaped, tongued

*glumaceus -a -um*   with chaffy bracts, conspicuously glumed

*glutinosus -a -um*   sticky, viscous, glutinous

*Glyceria*   Sweet

*Glycine*   Sweet (the roots of some species)

*glyco-, glycy-*   sweet-tasting or -smelling

*Glycyrrhiza*   Sweet-root (the source of liquorice)

*glypto-*   cut-into-, carved-

*glyptostroboides*   resembling *Glyptostrobus*

*Glyptostrobus*   Carved-cone (appearance of female cones)

*Gnaphalium*   Soft-down (from a Greek name for a plant with felted leaves)

*Gnidia, gnidium*   the Greek name for *Daphne*, from Gnidus, Crete

*Godetia*   for C. H. Godet, Swiss botanist (1797–1879)

*Goldbachia*   for C. L. Goldbach, writer on Russian medicinal plants (1793–1824)

*gompho-*   nail-, bolt- or club-shaped

*Gompholobium*   Club-pod (the shape of the fruit)

*gongylodes*   roundish, knob-like, swollen, turnip-shaped

*gonio-, -gonus -a -um*   angled-, prominently angled-

*Goodyera*   for John Goodyer, English botanist (1592–1664)

*gorgoneus -a -um*   gorgon-like, resembling one of the snake-haired Gorgons of mythology

*gossypi*   cotton-plant-like, resembling *Gossypium*

*Gossypium*   Soft (from Arabic *goz*, a soft substance)

*gothicus -a -um*   from Gothland, Sweden

*gracilior*   more graceful

*gracilis -is -e*   slender, graceful

*gracillimus -a -um*   most slender, most graceful

*Graderia*   an anagram of *Gerardia*, for John Gerard

*graecizans*   becoming widespread

*graecus -a -um*   Grecian, Greek

*gramineus -a -um*   grassy, grass-like

*gramini-, graminis -is -e*   of grasses, grass-like

*grammatus -a -um*   marked with raised lines or stripes

*Grammitis*   Short-line (sori appear to join up like lines of writing at maturity)

*granadensis -is -e, granatensis -is -e*   either from Granada in Spain, or from Colombia, South America, formerly New Granada

*grandi-, grandis -is -e*   large, powerful, full-grown, showy, big *graniticus -a -um*   of granitic rocks, grained

*granulatus -a -um, granulosos -a -um*   as though covered with granules, tubercled, granulate

*graph-, graphys-*   marked with lines-, as though written on

*grat-, gratus -a -um*   pleasing, graceful

*gratianopolitanus -a -um* from Grenoble, France

*Gratiola*   Agreeableness (medicinal effect)

*gratissimus -a -um*   most pleasing or agreeable

*graveolens*   strong-smelling, heavily scented

*Grevillea*   for C. F. Greville, founder member of the Royal Horticultural Society (1749–1809)

*Grewia*   for Nehemiah Grew, plant anatomist (1641–1712)

*Grindelia*   for D. H. Grindel, Latvian botanist (1776–1836)

*Griselinia*   for F. Griselini, Italian botanist (1717–1783)

*griseus -a -um*   bluish- or pearl-grey

*Groenlandia*   for Johannes Groenland of Paris

*groenlandicus -a -um*   from Greenland

*grosse, grossi-, grossus -a -um*   very large, thick, coarse

*Grossularia*   from the French 'groseille', gooseberry

*grossularioides*   gooseberry-like, resembling *Grossularia*

*gruinus -a -um*   crane-like

*grumosus -a -um*   broken into grains, tubercled, granular

*guadalupensis -is -e*   from Guadalupe Island off lower California, USA

*Guaiacum*   from the South American name for the wood of life tree

*guajava*   South American Spanish name

*guianensis -is -e*   from Guiana, northern South America

*guineensis -is -e*   from West Africa (Guinea Coast)

*Guizotia*   for Fr. P. G. Guizot, historian (1787–1874)

*gummifer -era -erum*   producing gum

*gummosus a -um*   gummy

*Gunnera*   for J. E. Gunner, Swedish botanist and cleric (1711–1777)

*guttatus -a -um*   spotted, covered with small glandular dots

*Gymnadenia*   Naked-gland (exposed viscidia of pollen)

*gymnanthus -a -um*   naked-flowered

*gymno-*   naked-

*Gymnocarpium*   Naked-fruit (sori lack indusia in oak fern)

*Gymnocladus*   Bare-branch (foliage mainly towards the ends of the branches)

*Gymnogramma*   Naked-line (the sori lack indusi)

*gyno-, -gynus -a -um*   relating to the ovary, female-, -carpelled

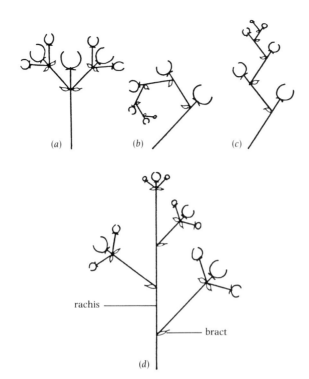

Fig. 3. Types of inflorescence which provide specific epithets:
(*a*), (*b*) and (*c*) are cymes, with the oldest flower in the centre or at
the apex of the inflorescence (e.g. *Saxifraga cymosa* Waldts. & Kit.);
(*b*) may have the three-dimensional form of a screw, or *bostryx*; (*c*)
may be coiled, or *scorpioid* (e.g. *Myosotis scorpioides* L.); (*d*) is a
raceme of cymes, or a *thyrse* (e.g. *Ceanothus thyrsiflorus* Eschw.)

*Gynura*  Female-tail (the stigma)
*Gypsophila*  Chalk-lover (the natural habitat)
*gyrans*  revolving, moving in circles
*gyro-*  bent-, twisted-
*gyrosus -a -um*  bent backwards and forwards

*Habenaria*  Thong (etymology uncertain)
*habr-, habro-*  soft-, beautiful-, delicate-

113

*hadriaticus -a -um*   from the shores of the Adriatic Sea

*haema-, haemalus -a -um, haemorrhoidalis -is -e, haematodes*
   blood-red, the colour of blood

*Haemanthus*   Blood-flower (the fireball lilies)

*haemanthus -a -um*   with blood-red flowers

*Hakea*   for Baron Hake, German horticulturalist (1745–1818)

*halepensis -is -e, halepicus -a -um*   from Aleppo, northern Syria

*Halesia*   for the Rev. Stephen Hales, writer on plants
   (1677–1761)

*halicacabum*   from an ancient Greek name

*halimi-, halimus -a -um*   orache-like, with silver-grey rounded
   leaves

*Halimione*   Daughter-of-the-sea

*Halimiocistus*   hybrids between *Halimium* and *Cistus*

*Halimium*   has leaves resembling those of *Atriplex halimus*

*halo-, halophilus -a -um*   salt-loving (the habitat)

*hama-*   together with-

*Hamamelis*   Greek name for a tree with pear-shaped fruits

*hamatus -a -um, hamosus -a -um*   hooked at the tip, hooked

*Hammarbya*   for Linnaeus who had a house at Hammarby in
   Sweden

*hamulatus -a -um*   having a small hook, clawed, talonned

*hamulosus -a -um*   covered with little hooks

*haplo-*   simple-, single-

*Haplopappus*   Single-down (its one-whorled pappus)

*harmalus -a -um*   adapting, responsive, sensitive

*Harpagophytum*   Grapple-plant (the fruit is covered with barbed
   spines)

*harpe-*   sickle-

*harpophyllus -a -um*   with sickle-shaped leaves

*hastati- hastatus -a -um*   formed like an arrow-head, spear-
   shaped (see Fig. 6(*a*)), hastate

*hastifer*   bearing a spear

*Hebe*   Greek goddess of youth, daughter of Jupiter

*hebe-*   pubescent- (youthful-)

*hebecarpus -a -um*   pubescent-fruited

*hebecaulis -is -e*   slothful-stemmed (prostrate)

*hebriacus -a -um*   Hebrew

*hecisto-*   viper-like-

*Hedera*   the Latin name for ivy

*hederaceus -a -um, hederi-*   ivy-like, resembling *Hedera* (usually in
   the leaf-shape)

*Hedychium(on)*   Sweet-snow (fragrant white flowers)

*Hedypnois*   Sweet-sleep
*hedys*   sweet, of pleasant taste or smell
*Hedysarum*   a name used by Dioscorides
*helena*   from Helenendorf, Transcaucasia
*Helenium*   for Helen of Troy (a name used by the Greeks for another plant)
*Heleocharis*   Marsh-favour (*Eleocharis*)
*heli-, helio-*   sun-
*Helianthemum*   Sun-flower
*Helianthus*   Sun-flower
*Helichrysum*   Golden-sun
*Heliconia*   for Mt Helicon, Greece, sacred to the Muses of mythology
*Helictotrichon(um)*   Twisted-hair (the awns)
*helioscopius -a -um*   sun-observing, sun-watching (the flowers track the sun's course)
*Heliotropium*   Turn-with-the-sun
*helix*   ancient Greek name for twining plants
*Helleborus*   Poison-food (the ancient Greek name for the medicinal *H. orientalis*)
*hellenicus -a -um*   from Greece, Grecian, Greek, Hellenic
*Helminthia* (*Helmintia*)   Worm (the elongate wrinkled fruits)
*helo-, helodes*   of bogs and marshes
*helodoxus -a -um*   marsh-beauty, glory of the marsh
*Helosciadium*   Marsh-umbel
*helveticus -a -um*   from Switzerland, Swiss
*helvolus -a -um*   pale yellowish-brown
*helvus -a -um*   dimly yellow, honey-coloured
*Helxine*   a name used by Dioscorides formerly for pellitory
*Hemerocallis*   Day-beauty (the flowers are short-lived)
*hemi-*   half-
*hemidartus -a -um*   patchily covered with hair, half-flayed
*hemionitideus -a -um*   barren, like a mule
*Hemionitis*   Mule (non-flowering – fern)
*Hemizonia*   Half-embraced (the achenes)
*Hepatica*   Liver (signature of leaf or thallus shape as of use for liver complaints)
*hepta-*   seven-
*Heracleum*   Hercules' -healer (a name used by Theophrastus)
*herba-barona*   fool's-herb (of the dunce or common man)
*herba-venti*   wind-herb (of the steppes)
*herbaceus -a -um*   not woody, low-growing, herbaceous
*hercoglossus -a -um*   with a coiled tongue

115

*hercynicus -a -um*  from the Harz mountains, mid Germany
*hermaeus -a -um*  from Mt Hermes, Greece
*Herminium*  Buttress (the pillar-like tubers)
*Hermodactylus*  Hermes'-fingers
*Herniaria*  Rupture-wort (former medicinal use)
*herpeticus -a -um*  ringworm-like
*Hesperis*  Evening (Theophrastus' name for evening-flowering)
*hespero-, hesperius -a -um*  western-, evening-
*heter-, hetero*  varying-, differing, diversely-
*Heteranthera*  Differing-anthers (has one large and two small)
*heteronema*  diverse-stemmed
*heterophyllus -a -um*  diversely-leaved
*Heteropogon*  Varying-beard (the twisting awns)
*Heuchera*  for J. H. Heucher, German professor of medicine (1677–1747)
*hex-, hexa-*  six-
*hexagonus -a -um*  six-angled
*hexandrus -a -um*  six-stamened
*hians*  gaping
*hibernalis -is -e*  of winter (flowering or leafing)
*hibernicus -a -um*  from Ireland (Hibernia), Irish
*hibernus -a -um*  flowering or green in winter, Irish
*Hibiscus*  an old Greek name for mallow
*hiemalis -is -e*  of winter
*Hieracium*  Hawkweed (Dioscorides' name for the supposed use of by hawks to give them acute sight)
*Hierochloe*  Holy-grass
*hierochunticus -a -um*  from the classical name for Jericho (*Anastatica h.* is the rose of Jericho)
*hieroglyphicus -a -um*  marked as if with signs
*Himantoglossum*  Strap-tongue (the narrow lip)
*Hippeastrum*  Knight-star
*hippo-*  horse-
*Hippocastanum*  Horse-chestnut
*Hippocrepis*  Horse-shoe (the shape of the fruit)
*hippomarathrum*  horse-fennel, Dioscorides' name for an arcadian plant which caused madness in horses
*Hippophae*  Horse-killer (used by Theophrastus for a prickly spurge)
*Hippuris*  Horse-tail
*hircinus -a -um*  of goats, smelling of a male goat
*Hirschfeldia*  for C. C. L. Hirschfeld, Austrian botanist
*hirsutissimus -a -um*  very hairy, hairiest

*hirsutulus -a -um, hirtellus -a -um, hirtulus -a -um*   somewhat hairy

*hirsutus -a -um*   rough-haired, hairy

*hirti-, hirtus -a -um*   hairy, shaggy-hairy

*hirundinaceus -a -um, hirundinarius -a -um*   pertaining to swallows

*hispalensis -is -e*   from Seville, southern Spain

*hispanicus -a -um*   from Spain, Spanish, Hispanic

*hispi-, hispidulus -a -um*   bristly, with stiff hairs

*Histiopteris*   Web-fern (the frond of Bat-wing fern)

*histrio-*   of varied colouring, theatrical

*histrionicus -a -um*   of actors, of the stage

*histrix*   showy, theatrical

*Hoheria*   from a Maori name

*Holcus*   Millet (the Greek name)

*hollandicus -a -um*   from either northern New Guinea or Holland

*Holmskjoldia*   for Theodore Holmskjold, Danish botanist (1732–1794)

*holo-*   completely-, entirely-, entire-

*Holoschoenus*   a name used by Theophrastus

*holosericus -a -um*   completly wrapped in silk

*Holosteum, holostea*   Whole-bone (an ancient Greek name for a chickweed-like plant)

*homal-, homalo*   smooth-, flat-

*Homalanthus*   Like-a-flower (the coloration of older leaves)

*Homalocephala*   Flat-head (the tops of the flowers)

*Homaria*   I-meet-together (the fused filaments)

*homo-, homoio-*   similar-, not varying-, agreeing with-, uniformly-

*Homogyne*   Uniform-female (the styles of neuter and female florets are not different)

*homolepis -is -e*   uniformly covered with scales

*hondensis -is -e*   from Hondo Island, Japan

*Honkenya*   for G. A. Honkeny, German botanist (1724–1805)

*hookeri, hookerianus -a -um*   for either Sir W. J. Hooker or his son Sir J. D. Hooker, both directors of Kew

*Hordelymus*   Barley-lime-grass

*Hordeum*   Latin name for barley

*horizontalis -is -e*   flat on the ground, spreading horizontally

*horminoides*   clary-like, resembling *Horminum*

*Horminum*   Exciter (the Greek name for sage used as an aphrodisiac)

*hormo-*   chain-, necklace-

*Hornungia*   for E. G. Hornung, German writer (1795–1862)

*horridus -a -um*   very thorny, rough, horridly armed

*hortensis -is -e, hortorum, hortulanus -a -um*   of gardens, cultivated

*hortulanorum*   of gardeners

*Hosta*   for N. T. Host, physician (1761–1834)

*Hottonia*   for Peter Hotton, Swedish botanist (1648–1709)

*Howea (Howeia)*   from the Lord Howe Islands, East of Australia

*hugonis*   for Fr Hugh Scallon, collector in West China

*humifusus -a -um*   spreading over the ground, sprawling

*humili-*   hop-

*humilis -is -e*   low-growing, smaller than most of its kind

*Humulus*   from the Slavic-German 'chmeli'

*hungaricus -a -um*   from Hungary, Hungarian

*hupehensis -is -e*   from Hupeh province, China

*Hura*   from a South American name

*Hutchinsia*   for Miss Hutchins, Irish cryptogamic botanist (1785–1815)

*Hyacinthus*   Homer's name for the flower which sprang from the blood of Hyakinthos, or from an earlier Thraco-pelasgian word for the blue colour of water

*hyacinthus -a -um, hyacinthinus -a -um*   dark purplish-blue, resembling *Hyacinthus*

*hyalinus -a -um*   nearly transparent, hyaline

*hybernalis -is -e, hybernus -a -um*   of winter

*hybridus -a -um*   bastard, mongrel, cross-bred, hybrid

*Hydrangea*   Water-vessel (shape of the capsules)

*Hydrilla*   Water-serpent

*hydro-*   water-, of water-

*Hydrocharis*   Water-beauty

*Hydrochloa*   Water-grass

*Hydrocotyle*   Water-cup

*hydrolapathum*   a name in Pliny for a water dock

*hydropiper*   water pepper

*hyemalis -is -e*   winter, of winter (flowering)

*hylaeus -a -um, hylo-*   of woods, of forests

*hylophilus -a -um*   wood-loving

*hymen-, hymeno-*   membrane-, membranous-

*Hymenanthera*   Membranous-stamen

*Hymenophyllum*   Membranous-leaf (filmy ferns)

*Hyoscyamus*   Hog-bean (a derogatory name by Dioscorides)

*Hyoseris*   Pig-salad (swine's succory)

*hypanicus -a -um*   from the region of the Hypanis River, Sarmatia
*Hyparrhenia*   Male-beneath (the arrangement of the spikelets)
*hyper-*   above-, over-
*hyperboreus -a -um*   of the far north
*Hypericum*   Above-pictures (early use over shrines to repel evil spirits)
*Hyphaene*   Network (the fibres in the fruit wall)
*hypnoides*   moss-like, resembling *Hypnum*
*hypo-*   under-, beneath-
*Hypochoeris*   a name used by Theophrastus
*hypochondriacus -a -um*   sombre, melancholy (colour)
*hypochrysus -a -um*   golden underside, golden beneath
*hypogaeus -a -um*   underground, subterrananean
*Hypolepis*   Under-scale (the protected sori)
*hypoleucus -a -um*   whitish, pale
*hypophegeus -a -um*   from beneath beech trees
*hypopithys*, *hypopitys*   growing under pine trees
*hyrcanus -a -um*   from the Caspian Sea area
*hyssopi-*   hyssop-like, resembling *Hyssopus*
*Hyssopus*   from a Semitic word 'ezob'
*hystri-*, *hystrix*   porcupine-like (the spiny corm of *Isoetes*)

*ianthinus -a -um*, *ianthus -a -um*   bluish-purple, violet-coloured
*iaponicus -a -um*   see *japonicus*
*-ias*   -much resembling
*ibericus -a -um*   either from Spain and Portugal (Iberia) or from the Georgian Caucasus
*iberideus -a -um*   from the Iberian peninsula
*iberidi-*   *Iberis*-like
*Iberis*   Dioscorides' name for an Iberian plant
*-ibilis -is -e*   able-, capable of-
*Icacina*   Icaco-like, resembling *Chrysobalanus icaco* (coco-plum)
*-icans*   -becoming, -resembling
*-icolus -a -um*   -of, -dwelling in
*icos-*   twenty-
*icosandrus -a -um*   twenty-stamened
*ictericus -a -um*   yellowed, jaundiced
*idaeus -a -um*   from Mount Ida in Crete, or Mount Ida in north-west Turkey
*-ides* (*-oides*)   -resembling, -similar to, -like
*Idesia*   for E. Y. Ides, Dutch explorer in China

*-idius -a -um*   -resembling
*idoneus -a -um*   worthy, apt, suitable
*Ifloga*   an anagram of *Filago*
*ignescens, igneus -a -um*   fiery-red
*il-, im-, in-*   in-, into-, for-, contrary-, contrarywise-
*Ilex*   the Latin name for the cork-oak (*Quercus ilex*)
*ilici-, ilicinus -a -um*   holly-, *Ilex-*
*-ilis -is -e*   -able, -having, -like, -resembling
*illecebrosus -a -um*   alluring, enticing, charming
*Illecebrum*   Charm (a name in Pliny)
*illinitus -a -um*   smeared, smudged
*-illius -a -um*   -lesser (a diminutive ending)
*illustratus -a -um*   pictured, painted, as if painted upon
*illustris -is -e*   brilliant, bright, clear
*illyricus -a -um*   from western Yugoslavia (Illyria)
*ilvensis -is -e*   from the Isle of Elba, or the river Elbe
*Ilysanthes*   Mud-flower
*imbecillis -is -e, imbecillus -a -um*   feeble, weak
*imberbis -is -e*   without hair, unbearded
*imbricans, imbricatus -a -um*   overlapping (leaves, bracts, scales)
   imbricate
*immaculatus -a -um*   unblemished, without spots, immaculate
*immarginatus -a -um*   without a rim or border
*immersus -a -um*   growing underwater
*impari-*   unpaired-, unequal-
*Impatiens*   Impatient (touch-sensitive fruits)
*impeditus -a -um*   tangled, hard to penetrate, impeding
*Imperata*   for the Italian botanist Imperato
*imperator, imperatoria*   emperor, ruler, master
*imperatricis*   for the Empress Josephine
*imperialis -is -e*   very noble, imperial
*implexus -a -um*   tangled
*impolitus -a -um*   dull, not shining, opaque
*imponens*   deceptive
*impressus -a -um*   sunken, impressed (e.g. leaf-veins), flattened
*impudicus -a -um*   lewd, shameless, impudent
*in-*   not-
*inaequalis -is -e*   unequal-sided
*inaequidens*   with unequal teeth, not equally toothed
*inapertus -a -um*   without an opening, closed, not open
*inatophyllus -a -um*   thong-leaved
*incanescens*   turning grey, becoming hoary
*incanus -a -um*   quite grey, hoary-white

*incarnatus -a -um*   made of flesh, flesh-coloured
*Incarvillea*   for Pierre d'Incarville, correspondent of B. de Jussieu
   from China (1706–1757)
*incertus -a -um*   doubtful, uncertain
*incisi-, incisus -a -um*   sharply and deeply cut into
*incomparabilis -is -e*   beyond compare, incomparable
*incomptus -a -um*   unadorned
*inconspicuus -a -um*   small
*incubaceus -a -um*   lying close to the ground
*incurvus -a -um*   inflexed, incurved
*indicus -a -um*   from India or, loosely, from the Orient
*Indigofera*   Indigo-bearer (source of blue dyes)
*induratus -a -um*   hard, indurate (usually of an outer surface)
*inebrians*   able to intoxicate, inebriating
*inermis -is -e*   without spines or thorns, unarmed
*-ineus -a -um*   -ish, -like
*inexpectans*   not expected (found where not expected)
*infaustus -a -um*   unfortunate
*infectorius -a -um*   dyed, used for dying
*infestus -a -um*   troublesome, hostile, dangerous
*infirmus -a -um*   weak, feeble
*inflatus -a -um*   swollen, inflated
*inflexus -a -um*   bent or curved inwards, inflexed
*infortunatus -a -um*   unfortunate (poisonous)
*infra-*   below-
*infractus -a -um*   curved inwards
*infundibuliformis -is -e*   trumpet-shaped, funnel-shaped
*ingens*   huge, enormous
*innatus -a -um*   natural, inborn, innate
*innominatus -a -um*   not named, unnamed
*innoxius -a -um*   without prickles, harmless
*inodorus -a -um*   without smell, scentless
*inominatus -a -um*   unlucky, inauspicious
*inophyllus -a -um*   fibrous-leaved
*inopinatus -a -um, inopinus -a -um*   surprising, unexpected
*inops*   deficient, poor
*inornatus -a -um*   without ornament, unadorned
*inquillinus -a -um*   introduced
*inquinans*   turning brown, staining, discolouring
*inscriptus -a -um*   as though written upon, inscribed
*insectifer -era -erum*   bearing insects (mimetic fly orchid)
*insertus -a -um*   inserted (the scattered inflorescences)
*insignis -is -e*   remarkable, decorative, striking

*insiticius -a -um, insititius -a -um, insitivus -a -um*   grafted

*insubricus -a -um*   from the Lapontine Alps (Insubria) between Lake Maggiore and Lake Lucerne

*insulanus -a -um, insularis -is -e*   growing on islands, insular

*intactus -a -um*   unopened, untouched (the flowers)

*integer -era -erum, integerrimus -a -um, integri-*   undivided, entire, intact

*integrifolius -a -um*   with entire leaves

*inter-*   between-

*interjectus -a -um*   intermediate in form, interposed (between two other species)

*intermedius -a -um*   between extremes, intermediate

*interruptus -a -um*   with scattered leaves or flowers

*intertextus -a -um*   interwoven

*intortus -a -um*   twisted

*intra-*   within-

*intricatus -a -um*   entangled

*introrsus -a -um*   facing inwards, turned towards the axis, introrse

*intumescens*   swollen

*intybus*   from a name in Virgil for wild chicory or endive

*Inula*   a name in Pliny for elecampane

*inundatus -a -um*   of marshes or places which flood, flooded

*-inus -a -um*   -ish, -like, -from

*invenustus -a -um*   lacking charm, unattractive

*inversus -a -um*   turned over, inverted

*involucratus -a -um*   surrounded with bracts, involucrate, with an involucre (the flowers)

*involutus -a -um*   obscured, rolled inwards, involute

*iodes*   violet-like, resembling *Viola*

*ioensis -is -e*   from Iowa, USA

*ion-, iono-*   violet-

*-ion*   -occurring

*ionantherus -a -um, ionanthes*   violet-flowered

*ionanthus -a -um*   with violet-coloured flowers

*ionicus -a -um*   from the Ionian Islands, Greece

*Ipheion*   a name used by Theophrastus

*iricus -a -um*   from Ireland, Irish

*iridescens*   iridescent

*iridi-*   *Iris*-like

*irio*   an ancient Latin name for a cruciferous plant

*Iris*   the name of the mythical goddess of the rainbow

*irrigatus -a -um*   of wet places, flooded

*irriguus -a -um*   watered (has clammy hairs)

*irritans*   causing irritation

*irroratus -a -um*   bedewed, dewy

*isabellinus -a -um*   yellowish, tawny

*isandrus -a -um*   equal-stamened, with equal stamens

*Isatis*   the name used by Hippocrates for woad

*Ischaemum*   Blood-stopper (a name in Pliny for its styptic property)

*-iscus -a -um*   -lesser (diminutive ending)

*islandicus -a -um*   from Iceland, Icelandic

*Isnardia*   for A-T. D. d'Isnard of Paris (1663–1743)

*iso-*   equal-

*Isoetes*   Equal-to-a-year (green throughout)

*Isolepis*   Equal-scales (the glumes)

*Isoloma*   Equal-lobes (of the perianth)

*-issimus -a -um*   -est, -the best, -the most (superlative)

*istriacus -a -um*   from Istria, Yugoslavia

*italicus -a -um*   from Italy, Italian

*Itea*   the Greek name for a Willow

*iteophyllus -a -um*   willow-leaved

*-ites, -itis*   -closely resembling, -very much like

*-ium*   -lesser (diminutive ending)

*Iva*   an old name applied to various fragrant plants

*ivorensis -is -e*   from the Ivory Coast, West Africa

*Ixia*   Bird-lime (Theophrastus' name refers to the clammy sap of these variably coloured plants)

*Ixiolirion*   Ixia-lily (the superficial resemblance)

*ixocarpus -a -um*   sticky-fruited

*Ixora*   the name of a Malabar deity


*Jacaranda*   the Brazilian name for *J. cuspidifolia*

*jacea*   from the Spanish name for knapweed

*jackmanii*   for G. Jackman, plant breeder of Woking

*jacobaeus -a -um*   either for St James (Jacobus) or from Iago Island, Cap Verde

*Jacobinia*   from Jacobina, South America

*jalapa*   from a South American Indian name

*jambos*   from a Malaysian name

*januensis -is -e*   from Genoa, north Italy, Genoan

*japonicus -a -um (iaponicus -a -um)*   from Japan, Japanese

*Jasione*   Healer (from a Greek name for *Convolvulus*)

*jasminoides*   jasmine-like, resembling *Jasminum*

*Jasminum*   from the Persian name 'yasmin'

*Jatropha*   Physician's-food (medicinal use)
*javanicus -a -um*   from Java, Javanese
*jejunifolius -a -um*   insignificant-leaved
*jejunus -a -um*   barren, poor, meagre, small
*jezoensis -is -e*   from Jezo (Yezo), Hokkaido, Japan
*johannis -is -e*   from Port St John, South Africa
*juanensis -is -e*   from Genoa, north Italy, Genoan
*Jubaea*   for King Juba of Numidia
*jubatus -a -um*   maned (crested with awns)
*jucundus -a -um*   pleasing
*jugalis -is -e*   joined together, yoked
*Juglans*   Jupiter's-nut (in Pliny – *Jovis glans*)
*jujuba*   from an Arabic name for *Zizyphus jujuba*
*juliae*   for Julia Mlokosewitsch who discovered *Primula juliae*
*julibrissin*   from the Persian name for *Acacia julibrissin*
*juliformis -is -e*   downy
*junceus -a -um, juncei-, junci-*   rush-like, resembling *Juncus*
*Juncus*   Binder (classical Latin name refers to use for weaving and basketry)
*juniperinus -a -um*   juniper-like, resembling *Juniperus*
*Juniperus*   the Latin name
*Jussiaea, Jussieua*   for Bernard de Jussieu who made a major contribution to establishing the concept of the taxonomic species (1699–1777)
*Justicia*   for J. Justice, Scottish gardener

*kaido*   a Japanese name
*kaki*   from the Japanese name *kaki-no-ki*
*Kalanchoe*   from a Chinese name
*kali-*   either from the Persian for a carpet, or a reference to the ashes of saltworts being alkaline (alkali)
*Kalmia*   for Peter Kalm, a highly reputed student of Linnaeus (1716–1779)
*kalo-*   beautiful-
*Kalopanax*   Beautiful-*Panax*
*kamtschaticus -a -um*   from the Kamchatka peninsula, eastern USSR
*kansuensis -is -e*   from Kansu province, China
*Kentranthus*   Spur-flower (see *Centranthus*)
*kermesinus -a -um*   carmine-coloured, carmine
*Kerria*   for William Kerr, collector of Chinese plants at Kew
*kewensis -is -e*   of Kew Gardens
*khasianus -a -um*   from the Khasi Hills, Assam

*Kickxia*   for J. J. Kickx, Belgian botanist (1842–1887)
*Kigelia*   from the native Mozambique name for the sausage tree
*kirro-*   citron-coloured-
*kisso-*   ivy-, ivy-like-
*kiusianus -a -um*   from Kiushiu, south Japan
*Knautia*   for Christian Knaut, German botanist (1654–1716)
*Kniphofia*   for J. H. Kniphof, German botanist (1704–1763)
*Kobresia* (*Cobresia*)   for Carl von Cobres, Austrian botanist
   (1747–1823)
*kobus*   a Japanese name
*Koeleria*   for L. Koeler, German botanist
*Koeningia* (*Koenigia*, *Koeniga*)   for J. G. König, student of
   Linnaeus, botanist in India (1728–1785)
*Kohlrauschia*   for H. Kohlrausch, German botanist
*kolomicta*   from a vernacular name from Amur, Eastern USSR
*koreanus -a -um*, *koraiensis -is -e*   from Korea, Korean
*kousa*   a Japanese name for a *Cornus* species
*kurroo*   from a Himalayan name

*labiatus -a -um*   lip-shaped, lipped, labiate
*labilis -is -e*   unstable, labile
*labiosus -a -um*   conspicuously lipped
*Laburnum*   a name in Pliny
*-lacca*   -resin
*lacciferus -a -um*   producing a milky juice
*lacer*, *lacerus -era*, *-erum*, *laceratus -a -um*   torn into a fringe, as
   if finely cut into
*Lachenalia*   for W. de Lachenal (de la Chenal), Swiss botanist
   (1763–1800)
*lachno-*   downy-, woolly-
*laciniatus -a -um*, *laciniosus -a -um*   jagged, unevenly cut, slashed
   (see Fig. 4(*f*))
*lacistophyllus -a -um*   having torn leaves
*lacrimans*   causing tears, weeping
*lacryma-jobi*   Job's-tears (shape and colour of fruit)
*lactescens*   having lac or milk
*lacteus -a -um*, *lact-*, *lacti-*   milk-coloured, milk-white
*lactifer -era -erum*   producing a milky juice
*Lactuca*   the Latin name (has a milky juice)
*lacunosus -a -um*   with gaps, furrows, pits or deep holes
*lacuster*, *lacustris -is -e*   of lakes or ponds
*ladanifer -era -erum*   bearing ladanum (the resin called myrrh)
*laetevirens*   bright-green

*laeti-, laetis -is -e, laetus -a -um*  pleasing, vivid, bright
*laevi-, laevigatus -a -um, laevis -is -e*  polished, not rough, smooth
*lag-, lago-*  hare's-
*lagaro-, lagaros-*  lanky, long, narrow, thin
*Lagenaria*  Flask (the Bottle-gourd fruit)
*lagenarius*  of a bottle or flask
*Lagerstroemia*  for Magnus von Lagerström of Göteborg, friend of
    Linnaeus
*lagopinus -a -um*  hare's-foot-like
*lagopus*  hare's foot
*Lagurus*  Hare's-tail (the inflorescence)
*Lamarckia (Lamarkia)*  for Jean Baptiste Antoine Pierre Monnet
    de Lamarck, evolutionist (1744–1829)
*lamellatus -a -um*  layered, lamellate
*lamii-*  deadnettle-like, resembling *Lamium*
*Lamiopsis*  looking like *Lamium*
*Lamium*  Gullet (the name in Pliny refers to the shape of the
    corolla-tube)
*lampr-, lampro-*  shining-, glossy-
*Lampranthus*  Shining-flower
*lanatus -a -um*  woolly
*lanceolatus -a -um, lanci-*  narrowed and tapered at both ends,
    lanceolate
*lanceus -a -um*  spear-shaped
*landra*  from the Latin name for a radish
*langleyensis -is -e*  from Veitch's Langley Nursery, England
*lani-, laniger -era -erum, lanosus -a -um, lanuginosus -a -um*
    softly-hairy, woolly or cottony
*Lantana*  an old name for *Viburnum*
*lapathi-*  sorrel-like-, dock-like-
*Lapeirousia (Lapeyrousia)*  for J. F. G. de la Peyrouse, French
    circumnavigator (1741–1788)
*lapidius -a -um*  hard, stony
*lappa, lappaceus -a -um*  bearing buds, bud-like
*lapponicus -a -um, lapponus -a -um*  from Lapland, of the Lapps
*lappulus -a -um*  with small burs (the nutlets)
*Lapsana (Lampsana)*  Purge (Dioscorides' name for a salad plant)
*larici-, laricinus -a -um*  larch-like, resembling *Larix*
*laricio*  the Italian name for several pines
*Larix*  Dioscorides' name for a larch
*Laser*  a Latin name for several umbellifers
*lasi-, lasio-*  shaggy-, woolly-
*lasiolaenus -a -um*  shaggy-cloaked, woolly-coated

*Lastrea*   for C. J. L. deLastre, French botanical writer
  (1792–1859)

*latebrosus -a -um*   of dark or shady places

*lateralis -is -e, lateri-*   on the side, laterally-

*latericius -a -um, lateritius -a -um*   brick-red

*Lathraea*   Hidden (fairly inconspicuous parasites)

*lathyris*   the Greek name for a kind of spurge

*Lathyrus*   the ancient name for the chickling pea

*lati-, latus -a -um*   broad, wide

*latifrons*   with broad fronds

*latipes*   broad-stalked, thick-stemmed

*latiusculus -a -um*   somewhat broad

*latobrigorum*   of the Rhinelands

*laudatus -a -um*   praised, worthy, lauded

*laureola*   of garlands (use of *Daphne laureola*)

*lauri-*   laurel-

*lauricatus -a -um*   wreathed, resembling laurel or bay

*laurinus -a -um*   laurel-like

*laurocerasus*   laurel-cherry (cherry-laurel)

*Laurus*   the Latin name for laurel or bay

*Laurustinus*   Laurel-like-*Tinus*

*lautus -a -um*   washed

*lavandulae-*   lavender-, *Lavandula-*

*Lavatera*   for the brothers Lavater, Swiss naturalists

*laxi-, laxus -a -um*   open, loose, not crowded, distant, lax

*lazicus -a -um*   from north-east Turkey (Lazistan)

*lecano-*   basin-

*Lecythis*   Oil-jar (the shape of the fruit)

*Ledum*   an old Greek name for a rockrose

*Leersia*   for J. D. Leers, German botanist (1727–1774)

*legionensis -is -e*   from León, Spain

*Legousia*   etymology uncertain

*leio-*   smooth-

*Lemna*   Theophrastus' name for a water-plant

*lendiger -era -erum*   nit-bearing (the appearance of the spikelets)

*Lens*   the classical name for the lentil

*Lentibularia*   usually regarded as referring to the lentil (*lens*)
  shaped bladders

*lenticularis -is -e*   lens-shaped, bi-convex

*lenticulatus -a -um*   with conspicuous lenticels on the bark

*lentiformis -is -e*   lens-shaped, bi-convex

*lentiginosus -a -um*   freckled, mottled

*lentus -a -um*   tough, pliable

127

*leodensis -is -e*   from Liège, Belgium
*leonensis -is -e*   from Sierra Leone, West Africa
*leonis -is -e*   toothed or coloured like a lion
*Leonotis*   Lion's-ear
*leonto-*   lion's-
*Leontodon*   Lion's-tooth
*Leonurus*   Lion's-tail
*Lepidium*   Little-scale (Dioscorides' name for a cress refers to the fruit)
*lepido-*   flaky-, scaly-
*Lepidotis*   Scaly
*lepidotus -a -um*   scurfy, scaly
*lepidus -a -um*   neat, elegant, graceful
*-lepis*   -scaly, -scaled
*Lepiurus*   Scale-tail (the inflorescence of sea hard grass, cf. *Pholiurus*)
*leporinus -a -um*   hare-like
*lept-*   slender-, hare-like-
*lepta-, lepto-*   slender-, weak-, thin-, small-, delicate-
*leptochilus -a -um*   with a slender lip
*leptophyllus -a -um*   slender-leaved
*Lepturus*   Hare's-tail
*Lespedeza*   for V. M. de Lespedez, Spanish politician in Florida, USA
*leuc-, leuco-*   white-
*Leucadendron*   White-tree
*Leucanthemum*   White-flower (Dioscorides' name)
*leuce*   a name for the white poplar
*Leucojum*   White-violet (Hippocrates' name for a snowflake)
*Leocorchis*   White-orchid
*levigatus -a -um*   smooth, polished
*levis -is -e*   smooth, not rough
*Levisticum*   Alleviator (the Latin equivalent of the Dioscorides' Greek name *Ligusticum*)
*Leycesteria*   for Wm. Leycester, horticulturalist in Bengal *c.* 1820
*libanensis -is -e, libanoticus -a -um*   from Mount Lebanon
*libani*   from the Lebanon, Lebanese
*libanotis -is -e*   from Mount Lebanon or of incense
*libericus -a -um*   from Liberia, West Africa
*libero-*   bark-
*liburnicus -a -um*   from Croatia (Liburnia) on the Adriatic
*libycus -a -um*   from Libya, Libyan

*lignescens*   turning woody

*lignosus -a -um*   woody

*ligtu*   from a Chilean name

*Ligularia*   Strap (the shape of the ray florets)

*ligularis -is -e, ligulatus -a -um*   strap-shaped, ligulate

*Ligusticum*   Dioscorides' name for a plant from Liguria, north-east Italy

*ligustrinus -a -um*   privet-like, resembling *Ligustrum*

*Ligustrum*   Binder (a name used in Virgil)

*lilacinus -a -um*   lilac-coloured, lilac-like

*lili-*   lily-

*liliaceus -a -um*   lily-like, resembling *Lilium*

*liliago*   silvery

*Lilium*   the name in Virgil

*lilliputianus -a -um*   of very small growth, Lilliputian

*limaeus -a -um*   of stagnant waters

*limbatus -a -um*   bordered, with a margin

*limbo-*   border-, margin-

*limensis -is -e*   from Lima, Peru

*Limnanthemum*   Pond-flower (spreads over surface)

*Limnanthes*   Marsh-flower

*limno-*   marsh-, pool-, pond-

*limnophilus -a -um*   marsh-loving

*limon*   the Persian name for *Citrus* fruits

*Limonium*   Dioscorides' name for a meadow plant

*Limosella*   Muddy

*limosus -a -um*   muddy, slimy, of muddy places

*lin-, linarii-, lini-*   flax-

*Linaria*   Flax-like (appearance of some species)

*linearis -is -e*   narrow and parallel-sided (usually the leaves)

*lineatus -a -um*   marked with lines (usually parallel and coloured)

*lingularis -is -e, lingulatus -a -um, linguus -a -um*   tongue-shaped (*Linguus* was a name in Pliny)

*linicolus -a -um*   of flax-fields

*linitus -a -um*   smeared

*Linnaea, linnaeanus -a -um, linnaei*   for Carl Linnaeus

*linosyris*   flax-coloured, an old generic name

*Linum*   the ancient Latin name for flax

*liolaenus -a -um*   smooth-cloaked, glabrous

*Liparis*   Greasy (the leaf-texture)

*Lippia*   for A. Lippi, French naturalist

*lirio-*   lily-white-

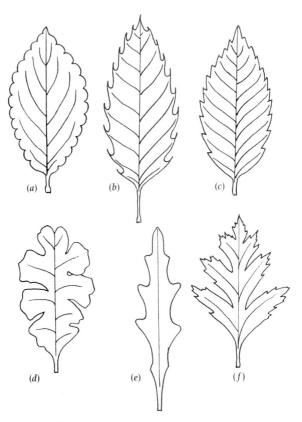

Fig. 4. Leaf-margin features which provide specific epithets:
(*a*) crenate (scalloped as in *Ardisia crenata* Sims); (*b*) dentate (toothed as in *Castanea dentata* Borkh.). This term has been used for a range of marginal tooth shapes; (*c*) serrate (saw-toothed as in *Zelkova serrata* (Thunb.) Makino); (*d*) lobate (lobed, as in *Quercus lobata* Née); (*e*) sinuate (wavy as in *Matthiola sinuata* (L.) R.Br.). This refers to 'in and out' waved margins, not 'up and down' or *undulate* waved margins; (*f*) laciniate (cut into angular segments as in *Crataegus laciniata* Ucria).

*Liriodendron*   Lily-tree (the showy flowers of the tulip tree)
*lisso-*   smooth-
*Listera*   for Dr M. Lister, pioneer palaeontologist (1638–1712)
*Litchi*   from the Chinese name
*literatus -a -um*   with the appearance of being written upon
*litho-*   stone-
*lithophilus -a -um*   living in stony places, stone-loving
*Lithops*   Stone-like (the appearance of stone-cacti)
*Lithospermum*   Stone-seed (the texture of the nutlets)
*lithuanicus -a -um*   from Lithuania, Lithuanian
*litigiosus -a -um*   disputed, contentious
*litoralis -is -e, littoralis -is -e, littorius -a -um*   of the sea-shore,
   growing by the shore
*Littorella*   Shore (the habitat)
*lividus -a -um*   lead-coloured, bluish-grey, leaden
*Livistonia*   for Patrick Murray, Lord Livingstone, whose garden
   formed the nucleus of the Edinburgh Royal Botanic Garden
*lizei*   for the Lizé Frères of Nantes, France
*Lloydia*   for Curtis G. Lloyd, American botanist (1859–1926)
*lobatus -a -um, lobi-, lobus -a -um*   with lobes, lobed (see
   Fig. 4 (*d*))
*lobbii, lobbianus -a -um*   for the brothers Thomas and William
   Lobb
*Lobelia*   for M. de l'Obel, renaissance pioneer of botany and
   herbalist to James I (1538–1616)
*lobius -a -um, -lobion*   pod, podded
*Lobularia*   Small-pod
*lochabrensis -is -e*   from Lochaber, Scotland
*lochmius -a -um*   coppice-dweller, of thickets
*locustus -a -um*   spikeleted
*loganobaccus -a -um*   loganberry, after its developer Judge J. H.
   Logan
*Loiseleuria*   for J. L. A. Loiseleur-Deslongchamps, French botanist
   (1774–1849)
*loliaceus -a -um*   resembling *Lolium*
*Lolium*   a name in Virgil for a weed grass
*-loma*   -fringe, -border
*Lomaria*   Border (the marginal sori)
*lonchitis -is -e, loncho-*   spear-shaped, lance-shaped (a name used
   by Dioscorides for a fern)
*longe-, longi-, longus -a -um*   long
*longipes*   long-stalked
*Lonicera*   for Adam Lonitzer, German botanist (1528–1586)

*lophanthus -a -um*   with crested flowers

*lopho-*   crest-, crested-

*Lophophora*   Crest-bearer (has tufts of glochidiate hairs)

*Loranthus*   Strap-flower (the shape of the 'petals')

*loratus -a -um*, *lori-*, *loro-*   strap-shaped

*lorifolium -a -um*   with long narrow leaves, strap-leaved

*Loroglossum*   Strap-tongue (the elongate lip)

*loti-*, *lotoides*   trefoil-like, resembling *Lotus*

*Lotus*   the ancient Greek name for various leguminous plants

*louisianus -a -um*   from Louisiana, USA

*loxo-*   oblique-

*lucens*, *lucidus -a -um*   glittering, shining, clear

*luciae*   for Madame Lucie Savatier

*luciliae*   for Lucile Boissier

*lucorum*   of woodland or woods

*ludovicianus -a -um*   from Louisiana, USA

*Ludwigia*   for C. G. Ludwig, German botanist (1709–1773)

*Luffa*   from the Arabic name

*Lunaria*   Moon (the shape and colour of the septum (or repla) of the fruit of honesty)

*lunatus -a -um*, *lunulatus -a -um*   half-moon-shaped, lunate

*Lupinus*   the ancient Latin name for the white lupin

*lupuli-*, *lupulinus -a -um*   hop-like, resembling *Humulus*

*Lupulus*   Wolf (the ancient Latin name for hop was a reference to its straggling habit on other plants – willow wolf)

*luridus -a -um*   sallow, dingy yellow or brown, wan, lurid

*Luronium*   Rafinesque's name for a water plantain

*lusitanicus -a -um*   from Portugal (Lusitania), Portuguese

*lutarius -a -um*   of muddy places, living on mud

*luteo-*, *luteus -a -um*   yellow

*luteolus -a -um*   yellowish

*lutescens*   turning yellow

*lutetianus -a -um*   from Paris (Lutetia), Parisian

*luxurians*   luxuriant

*Luzula*   an ancient name of obscure meaning

*Lychnis*   Lamp (the hairy leaves were used as wicks for oil lamps)

*lychnitis*   from a name in Pliny meaning of lamps

*Lycium*   the name of a thorn tree from Lycia

*lycius -a -um*   from Lycia, south-west Turkey

*lyco-*   wolf- (usually implying inferior or wild)

*Lycocarpus*   Wolf-fruit (clawed at the upper end)

*lycoctonum*   wolf-murder (poisonous wolf's-bane aconite)

*Lycopersicum(on)*   Wolf-peach (tomato)
*Lycopodium*   Wolf's-foot (clubmoss)
*Lycopsis*   Wolf-like (Dioscorides' derogatory name)
*Lycopus*   Wolf's-foot
*Lycoris*   for Marc Antony's mistress
*lydius -a -um*   from Lydia, south-west Turkey
*Lygodium*   Willow-like (the climbing fern's stems)
*lyratus -a -um*   lyre-shaped (rounded above with small lobes
   below – usually of leaves)
*lysi-, lysio-*   loose-
*Lysichiton(um)*   Loose-cloak (the deciduous spathe)
*Lysimachia*   Ending-strife (named after a Thracian king –
   loosestrife)
*Lythrum*   Black-blood (Dioscorides's name refers to the flower
   colour of some species)

*macedonicus -a -um*   from Macedonia, Macedonian
*macellus -a -um*   rather meagre, poorish
*macer -ra -rum*   meagre
*macilentus -a -um*   thin, lean
*macr-, macro-*   big-, large-, long-
*macrodus -a -um*   large-toothed
*macromeris -is -e*   with large parts
*macrorhizus -a -um*   large-rooted
*macrosiphon*   large-tubular, long-tubed
*macrurus -a -um*   long-tailed
*maculatus -a -um, maculi-, maculifer -era -erum*   spotted, blotched,
   bearing spots
*madagascariensis -is -e*   from Madagascar, Madagascan
*maderaspatanus -a -um, maderaspatensis -is -e*   from the Madras
   region of India
*maderensis -is -e*   from Madeira, Madeiran
*Madia*   from a Chilean name
*madrensis -is -e*   from the Sierra Madre, northern Mexico
*madritensis -is -e*   from Madrid, Spain
*maesiacus -a -um*   from the Bulgarian Serbian region once called
   Maesia
*magellanicus -a -um*   from the Straits of Magellan, South
   America
*magellensis -is -e*   from Monte Majella, Italy
*magni-, magno-, magnus -a -um*   large
*magnificus -a -um*   great, eminent, distinguished, magnificent
*magnifolius -a -um*   large-leaved

133

*Magnolia*   for Pierre Magnol of Montpelier (1638–1715)
*mahaleb*   an Arabic name
*Mahonia*   for B. M'Mahon, American horticulturalist
   (1775–1816)
*mai-, maj-*   May-
*Maianthemum*   May-flower (a May-flowering lily)
*majalis*   of the month of May (flowering time)
*majesticus -a -um*   majestic
*major -or -us*   larger, greater, bigger
*malabaricus -a -um*   from the Malabar coast, south India
*malaco-, malako-, malacoides*   soft, tender, weak, mucilaginous
*Malaxis*   Softener (soft leaves)
*Malcolmia* (*Malcomia*)   for Wm. Malcolm, English
   horticulturalist *c.* 1798
*maliformis -is -e*   apple-shaped
*mallophorus -a -um*   wool-bearing
*Malope*   a name for mallow in Pliny
*Malpighia*   for Marcello Malpighi, Italian naturalist
   (1628–1694)
*Malus*   the Latin name for an apple tree
*Malva*   Soft (the name in Pliny)
*malvaceus -a -um*   mallow-like, resembling *Malva*
*malvinus -a -um*   mauve, mallow-like
*mammaeformis -is -e, mammiformis -is -e*   shaped like a nipple
*Mammillaria* (*Mamillaria*)   Nipple (conspicuous tubercles)
*mammillatus -a -um, mammillaris -is -e, mammosus -a -um*
   having nipple-like structures, mammillate
*mandibularis -is -e*   jaw-like, having jaws
*Mandragora*   a Greek name derived from a Syrian one
*mandschuricus -a -um, mandshuricus -a -um*   from Manchuria,
   Manchurian
*Mangifera*   from the Hindu name for the fruit
*manicatus -a -um*   with long sleeves, with a felty covering which
   can be stripped off, manicate
*manipuliflorus -a -um*   with few-flowered clusters
*manipuranus -a -um*   from Manipur, India
*manriqueorum*   for Manrique de Lara, of the Manriques
*mantegazzianus -a -um*   for Paulo Mantegazzi, Italian traveller
*Manzanilla*   from the Spanish for a small apple, 'manzanita'
*Maranta*   for Bartolomea Maranti, Venetian botanist
*marcescens*   not putrefying, persisting, retaining dead leaves
*marckii*   for J. B. A. P. Monnet de la Marck; Lamarck
*margaritaceus -a -um, margaritus -a -um*   pearly, of pearls

*margaritiferus -a -um*   bearing pearl

*marginalis -is -e*   of the margins, marginal

*marginatus -a -um*   having a distinct margin (the leaves)

*marianus -a -um*   of St Mary, from Maryland, USA, or from the Sierra Morena

*marilandicus -a -um, marylandicus -a -um*   from the Maryland region, USA

*marinus -a -um*   growing by or in the sea, marine

*mariscus -a -um*   the name for a rush in Pliny

*maritimus -a -um*   growing by the sea, of the sea, maritime

*marmelos*   a Portuguese name

*marmoratus -a -um, marmoreus -a -um*   marbled

*maroccanus -a -um*   from Morocco, north-east Africa, Moroccan

*Marrubium*   the old Latin name

*Marsilea*   for Ludwig F. Marsigli, Italian patron of botany (1658–1730)

*martagon*   resembling a kind of Turkish turban

*Martia, Martiusia*   for K. F. P. von Martius, German botanist in Brazil (1794–1868)

*martinicensis -is -e*   from Martinique

*maru*   mastic

*marus -a -um*   glowing

*mas, maris, masculus -a -um*   bold, male

*massiliensis -is -e*   from Marseilles, France

*mastichinus -a -um*   gummy, mastic-like

*mastigophorus -a -um*   producing gum, gum-bearing

*Matricaria*   Mother-care (former medicinal use in treatment of uterine infections)

*matritensis -is -e*   from Madrid, Spain

*matronalis -is -e*   of married women (the Roman matronal festival was held on March 1st)

*matsudana*   for Sadahisa Matsudo, Japanese botanist (1857–1921)

*Matteuccia (Matteucia)*   for C. Matteucci, Italian physicist (1800–1868)

*Matthiola*   for Pierandrea A. G. Matthioli, Italian botanist (1500–1577)

*mauritanicus -a -um*   from Morocco or North Africa generally

*mauritianus -a -um*   from the island of Mauritius, Indian Ocean

*maurorum*   of the Moors, Moorish, of Mauritania

*maxillaris -is -e*   of jaws, resembling an insect's jaws

*maximus -a -um*   largest, greatest

*mays*   from the Mexican name for Indian corn

*Meconopsis*   Poppy-like

*Medicago*   from a Persian name for a grass

*medicus -a -um*   from Media (Iran), curative, medicinal

*medio-, medius -a -um*   middled-sized, in between, intermediate

*mediopictus -a -um*   with a coloured stripe down the centre-line (of a leaf)

*mediterraneus -a -um*   from the Mediterranean region, from well inland

*medullaris -is -e, medullosus -a -um, medullus -a -um*   pithy, soft-wooded with a large pith

*mega-, megalo-*   big-, great-, large-

*megalurus -a -um*   large-tailed

*megapotamicus -a -um*   of the big river, from the Rio Grande or Amazon river

*megaseifolius -a -um*   *Bergenia*-leaved (Megasea-leaved)

*mela-, melan-, melano-*   black-

*Melaleuca*   Black-and-white (the colours of the bark on trunk and branches)

*Melampyrum(on)*   Black-wheat (a name used by Theophrastus)

*melancholicus -a -um*   sad-looking, drooping, melancholy

*Melandrium*   Black-oak (the name used in Pliny)

*melanops*   black-eyed

*melanoxylon*   black-wooded

*Melastoma*   Black-mouth (the fruits stain)

*meleagris -is -e*   Greek name for Meleagris of Calydon, or chequered like a guinea fowl

*Melia*   from the Greek name for ash (the resemblance of the leaves)

*Melianthus*   Honey-flower

*Melica*   Honey-grass

*Melilotus*   Honey-lotus (Theophrastus' name refers to melilot's attractiveness to bees)

*Melissa*   Bee-keeper (named for the nymph who in mythology kept bees, and the plant's use in apiculture)

*melissophyllus -a -um*   balm-leaved, with *Melissa*-like leaves

*melitensis -is -e*   from Malta, Maltese

*Melittis*   Bee (bastard balm attracts bees)

*mellifer -era -erum*   honey-bearing

*mellitus -a -um*   darling, honey-sweet

*melo-*   melon-

*Melocactus*   melon-cactus (the shape)

*meloniformis -is -e*   like a ribbed-sphere, melon-shaped

*melongena*   apple-bearer (producing a tree-fruit, the egg plant)

*Melothria*   the Greek name for bryony
*membranaceus -a -um*   thin in texture, skin-like, membranous
*Mentha*   the name in Pliny
*mentorensis -is -e*   from Mentor, Ohio, USA
*Menyanthes*   Moon-flower (Theophrastus' name for
  Nymphoides)
*Menziesia*   for Archibald Menzies, English naturalist on the
  *Discovery* (1754–1842)
*Mercurialis -is -e*   named by Cato for Mercury, messenger of the
  Gods
*meridianus -a -um, meridionalis -is -e*   of noon, flowering at
  midday, southern
*Mertensia*   for F. C. Mertens, German botanist (1764–1831)
*-merus -a-um*   -partite, -divided into, -merous
*mes-, meso-*   middle-
*Mesembryanthemum*   Midday-flower (flowers open in full sun)
*mesoponticus -a -um*   from the middle sea (lakes of central
  Africa)
*mesopotamicus -a -um*   from between the rivers
*Mespilus*   Theophrastus' name for the medlar
*messaniensis -is -e*   from Messina, Italy
*messeniensis -is -e*   from Messenia, Morea, Greece
*met-, meta-*   amongst-, next to-, after-, behind-, later-
*metallicus -a -um*   lustrous, metallic in appearance
*Metasequoia*   Close-to-*Sequoia* (resemblance of the dawn
  redwood)
*methysticus -a -um*   intoxicating
*Metrosideros*   Pith-of-iron (the hard timber)
*Metroxylon*   Pith-wood (the large medulla)
*Meum* (*Meon*)   an old Greek name in Dioscorides
*mexicanus -a -um*   from Mexico, Mexican
*Mibora*   an Andansonian name of uncertain meaning
*micaceus -a -um*   from mica soils
*micr, micro-*   small-
*micranthus -a -um*   small-flowered
*Microcala*   Little-beauty
*microcarpus -a -um*   small-fruited
*microdasys*   small and hairy, with short shaggy hair
*microdon*   small-toothed
*microglochin*   small-point (the tip of the flowering axis)
*Microsisymbrium*   Little-*Sisymbrium*
*mikanioides*   resembling *Mikania* (climbing hemp-weed)
*miliaceus -a -um*   millet-like, pertaining to millet

*militaris -is -e*   soldier-like, resembling part of a uniform
*Milium*   the Latin name for a millet grass
*mille-*   a thousand- (usually means 'very many')
*millefolius -a -um*   thousand-leaved (much divided leaves of milfoil)
*mimetes*   mimicking
*Mimosa*   Mimic (the sensitivity of the leaves)
*Mimulus*   Ape-flower
*miniatus -a -um*   cinnabar-red, the colour of red lead
*minimus -a -um*   least, smallest
*minor -or -us*   smaller
*Minuartia*   for Juan Minuart, Spanish botanist (1693–1768)
*minutissimus -a -um*   extremely small, smallest
*minutus -a -um*   very small, minute
*mirabilis -is -e*   wonderful, extraordinary, astonishing
*mirandus -a -um*   extraordinary
*Misopates*   Reluctant to open
*missouriensis -is -e*   from Missouri, USA
*Mitella*   Little-mitre (the shape of the fruit)
*mitis -is -e*   gentle, mild, bland, not acid, without spines
*mitratus -a -um*   turbaned, mitred
*mitriformis -is -e*   mitre-shaped
*mixtus -a -um*   mixed
*modestus -a -um*   modest
*Moehringia*   for P. H. G. Möhring, German naturalist (1710–1792)
*Moenchia*   for Conrad Moench, German botanist (1744–1805)
*moesiacus -a -um*   from the Balkans (Moesia)
*moldavicus -a -um*   from the Danube area (Moldavia)
*molinae, Molinia*   for J. I. Molina, writer on Chilean plants (1740–1829)
*Molium*   Magic-garlic (after *Allium moly*)
*molle*   from Peruvian 'mulli' for *Schinus molle*
*molliaris -is -e*   supple, graceful, pleasant
*mollis -is -e*   softly-hairy, soft
*Mollugo*   Soft (a name in Pliny)
*moluccanus -a -um*   from Indonesia (the Moluccas)
*moly*   the Greek name of a magic herb
*mona, mono-*   one-, single-, alone-
*monadelphus -a -um*   in one group or bundle (stamens)
*monandrus -a -um*   one stamened, with a single stamen
*monensis -is -e*   from Anglesey or the Isle of Man, both formerly known as Mona

*Moneses*   One-product (the solitary flower)

*mongholicus -a -um, mongolicus -a -um*   from Mongolia, Monglian

*moniliformis -is -e*   necklace-like, like a string of beads

*monococcus -a -um*   one-fruited or -berried

*Monodora*   One-gift (the solitary flowers)

*Monotropa*   One-turn (the band at the top of the stem)

*monspeliensis -is -e, monspessulanus -a -um*   from Montpellier, south France

*monstrosus -a -um*   abnormal, monstrous

*montanus -a -um, monticolus -a -um*   of mountains, mountain-dweller

*Montbretia*   for A. F. E. C. de Montbret, botanist in Egypt (1781–1801)

*montevidensis -is -e*   from Montevideo, Uruguay

*Montia*   for G. L. Monti, Italian botanist (1712–1797)

*Moricandia*   for M. E. (Stephan) Moricand, Swiss botanist (1779–1854)

*morifolius -a -um*   mulberry-leaved, with *Morus*-like leaves

*morio*   madness

*-morphus -a -um*   -shaped, -formed

*morsus-ranae*   mouth of the frog (frog-bit)

*mortuiflumis -is -e*   of dead water, growing in stagnant water

*Morus*   the ancient Latin name for the mulberry

*moschatus -a -um*   musk-like, musky (scented)

*mosiacus -a -um*   parti-coloured, coloured like a mosaic

*moupinensis -is -e*   from Mupin, west China

*mucosus -a -um*   slimy

*mucronatus -a -um*   with a hard sharp-pointed tip, mucronate (see Fig. 7 (*b*))

*mugo*   an old name for the dwarf pine

*Mulgedium*   Milker (Cassini's name refers to the possession of latex as in *Lactuca*)

*multi-, multus -a -um*   many

*multicavus -a -um*   with many hollows, many-cavitied

*multiceps*   many-headed

*multijugus -a -um*   pinnate, with many pairs of leaflets

*mume*   from the Japanese 'ume'

*mundus -a -um*   clean, neat, elegant, handsome

*munitus -a -um*   armed, fortified

*muralis -is -e*   of walls, growing on walls

*muricatus -a -um*   rough with short superficial tubercles, muricate

*murinus -a -um*   mouse-grey, of mice

*murorum*   of walls

*Musa*   for Antonia Musa, physician to Emperor Augustus
   (63–14 BC)

*musaicus -a -um*   mottled like a mosaic, resembling *Musa*

*Muscari*   Musk-like (from the Turkish – fragrance)

*muscifer -era -erum*   fly-bearing (floral resemblance)

*muscipulus -a -um*   fly-catching

*muscivorus -a -um*   fly-eating

*muscoides*   fly-like, moss-like

*muscosus -a -um*   moss-like, mossy

*musi-*   banana-

*mutabilis -is -e*   changeable (in colour), mutable

*mutatus -a -um*   changed, altered

*muticus -a -um*   without a point, not pointed, blunt

*Mutisia (Mutisa)*   for J. C. B. Mutis y Bosio, Spanish discoverer of
   *Cinchona* (1732–1808)

*myagroides*   resembling *Myagrum*

*Myagrum*   Mouse-trap (Dioscorides' name)

*Mycelis*   de l'Obel's name has no clear meaning

*myiagrus -a -um*   fly-catching (sticky)

*Myosotis*   Mouse-ear (Dioscorides' name)

*Myosoton*   Mouse-ear (Dioscorides' name)

*Myosurus*   Mouse-tail (the fruiting receptacle)

*Myrica*   Fragrance (the ancient name for a tamarisk)

*myrio-*   numerous-, myriad-

*Myriophyllum*   Numerous-leaves (Dioscorides' name)

*Myristica*   Myrrh-fragrant

*myrmecophilus -a -um*   ant-loving (plants with special ant
   accommodations and associations)

*Myrrhis*   the ancient name for true myrrh

*myrsinites*   myrtle-like

*myrtus*   the Greek name for myrtle

*mystacinus -a -um*   moustached

*mysurensis -is -e*   from Mysore, India

*myuros*   mouse-tailed

*Myurus*   Mouse-tail (the fruiting receptacle)


*naevosus -a -um*   freckled, with mole-like blotches

*Naias, Najas*   Water-nymph (its habitat)

*nairobensis -is -e*   from Nairobi, Kenya

*namaquensis -is -e*   from Namaqualand, western South Africa

*nana-, nanae-, nani-, nano-, nanoe-, nanus -a -um*   dwarf

*nanellus -a -um*   very dwarf

*nannophyllus -a -um*   small-leaved
*napaeifolius -a -um (napeaefolius -a -um)   Napaea*-leaved
*napaulensis -is -e*   from Nepal, Nepalese
*napellus -a -um*   swollen, turnip-rooted, like a small turnip
*napi-*   turnip-
*Napoleonaea (Napoleona)*   for Napoleon Bonaparte
*Napus*   the name in Pliny for a turnip
*narbonensis -is -e*   from Narbonne, southern France
*Narcissus*   the name of a youth in Greek mythology who fell in
   love with his own reflection, torpid (the narcotic effect)
*Nardurus   Nardus*-tail (the narrow inflorescence)
*Nardus*   Spikenard-like
*narinosus -a -um*   broad-nosed
*Narthecium*   Little-rod (the stem)
*Nasturtium*   Nose-twist (the mustard-oil smell)
*natans*   floating, swimming
*Naumbergia*   for S. J. Naumberg, German botanist (1768–1799)
*nauseosus -a -um*   nauseating
*navicularis -is -e*   boat-shaped
*nebrodensis -is -e*   from Mt Nebrodi, Sicily
*nebulosus -a -um*   cloud-like, clouded, nebulous
*neglectus -a -um*   (formerly) overlooked, disregarded, neglected
*negundo*   from a Sanskrit name for a tree with leaves like box-
   elder
*nelumbo*   from a Sinhalese name
*-nema, nema-, nemato-*   thread-like-
*Nemesia*   a name used by Dioscorides for another plant
*Nemophila*   Glade-loving (woodland habitat)
*nemoralis -is -e, nemorosus -a -um, nemorum*   of woods, sylvan
*neo-*   new-
*neomontanus -a -um*   from Neuberg, Germany
*neopolitanus -a -um*   from Naples, neopolitan
*Neotinnea (Neotinea)*   New-*Tinnea* (for similarity to the genus
   named for three Dutch ladies who explored on the Nile)
*Neottia*   Nest-of-fledglings
*nepalensis -is -e*   from Nepal, Nepalese
*Nepenthes*   Euphoria (its reputed drug property)
*nepeta*   from Nepi, Italy
*nephr-, nephro-*   kidney-shaped-, kidney-
*Nephrodium*   Kidneys (the shape of the indusia of the sori)
*Nephrolepis*   Kidney-scale (the shape of the indusia of the sori)
*nericus -a -um*   from the province of Narke, Sweden
*nerii-*   oleander-like-, *Nerium*-

*neriifolius -a -um (nereifolius -a -um)* Nerium-leaved
*Nerium* the ancient Greek name for oleander
*nerterioides* resembling *Nertera* (bead plants)
*nervatus -a -um, nervis -is -e* nerved or veined
*nervosus -a -um* with conspicuous nerves or veins
*Neslia* for the French botanist, Nesles
*nesophilus -a -um* island-loving
*nessensis -is -e* from Loch Ness, Scotland
*-neurus -a -um* -nerved, -veined
*nevadensis -is -e* from Nevada or the Sierra Nevada, USA
*nicaensis -is -e* from Nice, south-east France or Nicaea,
   Bithynia, north-west Turkey
*Nicandra* for Nicander of Calophon, writer on plants (100 BC)
*Nicotiana* for Jean Nicot who introduced tobacco to France in
   the late sixteenth century
*nictitans* moving, blinking
*nidi-, nidus* nest, nest-like
*nidus-aves* bird's-nest (resemblance)
*Nigella* Blackish (the seed coats)
*niger -ra -rum* black
*nigericus -a -um* from Nigeria, West Africa
*nigrescens, nigri-, nigro-, nigricans* blackish, darkening, turning
   black
*nikoensis -is -e* from Nike, Japan
*niliacus -a -um* from the River Nile
*niloticus -a -um* from the Nile valley
*nipho-* snow-
*nipponicus -a -um* from Japan (Nippon), Japanese
*nitens, nitidi-, nitidus -a -um* glossy, with a polished surface,
   neat
*nivalis -is -e* snow-white, growing near snow
*niveus -a -um, nivosus -a -um* purest white, snow-white
*nobilis -is -e* famous, grand, noble
*nocti-* night-
*noctiflorus -a -um, nocturnus -a -um* night-flowering
*nodiflorus -a -um* flowering at the nodes
*nodosus -a -um* many-jointed, conspicuously jointed, knotty
*nodulosus -a -um* with swellings (on the roots), noduled
*Nolana* Small-bell
*noli-tangere* Touch-not (the ripe fruit ruptures on touch)
*noma-, nomo-* meadow-
*nominius -a -um* customary
*non-* not-, un-

*nonpictus -a -um*   of plain colour, not painted

*nonscriptus -a -um*   unmarked, not written upon

*nootkatensis -is -e, nutkatensis -is -e*   from Nootka Sound, British Columbia, Canada

*normalis -is -e*   representative of the genus, normal

*norvegicus -a -um*   from Norway, Norwegian

*notatus -a -um*   spotted

*notho-, nothos-, nothus -a -um*   false-, bastard-, spurious-

*Nothoscordum*   Bastard-garlic

*noti-, notio-*   southern-

*noto-*   the back-, surface-

*novae-caesareae* (*novi-caesareae*)   from New Jersey, USA

*novae-zelandiae*   from New Zealand

*noveboracensis -is -e*   from New York, USA

*novi-belgii* (*novi-belgae*)   from New York, USA

*novi-caesareae*   from New Jersey, USA

*nubicolus -a -um, nubigenus -a -um, nubilus -a -um*   of cloudy places

*nubicus -a -um*   from the Sudan (Nubia), north-east Africa

*nubilorum*   from high peaks, of clouds

*nucifer -era -erum*   nut-bearing

*nudatus -a -um, nudi-, nudus -a -um*   bare, naked

*nudicaulis -is -e*   naked-stemmed, leafless

*numidicus -a -um*   from Algeria (Numidia), Algerian

*nummularis -is -e*   circular, coin-like (the leaves)

*nummularius -a -um*   money-wort-like, resembling *Nummularia*

*Nuphar*   the Persian name for a water lily

*nutabilis -is -e*   sad-looking, drooping

*nutans*   drooping, nodding (the flowers)

*nutkanus -a -um*   as for *nootkatensis*

*nux-vomica*   with nuts causing vomiting (contain strychnine)

*-nychius -a -um*   -clawed

*nycticalus -a -um, nyctagineus -a -um*   night-flowering

*Nymphaea*   Nymphe (Theophrastus' name after one of the three water nymphs)

*Nymphoides*   resembling *Nymphaea*

*Nyssa*   Nyssa (another water nymph)


*ob-, oc-, of-, op-*   contrary-, inverted-, inversely-, against-

*obconicus -a -um*   like an inverted cone

*obcordatus -a -um*   inversely cordate (stalked at narrowed end of a heart-shaped leaf), obcordate

*obesus -a -um*   succulent, fat

143

*obfuscatus -a -um*  clouded over, confused

*Obione*  Daughter-of-the-Obi (a Siberian river)

*oblatus -a -um*  somewhat rounded at the ends, oval, oblate

*obliquus -a -um*  slanting, unequal-sided, oblique

*oblongatus -a -um, oblongi, oblongus -a -um*  elliptic with blunt ends

*obovatus -a -um*  egg-shaped in outline and with the narrow end lowermost

*obscurus -a -um*  dark, darkened, obscure

*obsoletus -a -um*  rudimentary

*obtectus -a -um*  covered over

*obtusatus -a -um, obtusi-, obtusus -a -um*  blunt, rounded, obtuse

*obtusior*  more obtuse (than the type)

*obvallaris -is -e, obvallatus -a -um*  walled around, enclosed

*obvolutus -a -um*  half-amplexicaule, with one leaf margin overlapping that of its neighbour

*occidentalis -is -e*  western, occidental

*occultus -a -um*  hidden

*oceanicus -a -um*  growing near the sea

*ocellatus -a -um*  eye-like, with a colour-spot bordered with another colour

*ochnaceus -a -um*  resembling *Ochna*

*ochraceus -a -um*  ochre-coloured, yellowish

*ochroleucus -a -um*  buff-coloured, yellowish-white

*ocimoides, ocymoides*  sweet basil-like, resembling *Ocimum*

*Ocimum*  the Greek name for an aromatic plant

*octa-, octo-*  eight-

*octandrus -a -um*  eight-stamened

*oculatus -a -um*  eyed, with an eye

*-odes*  -like, -resembling

*odessanus -a -um*  from Odessa, Black Sea area of USSR

*Odontites*  For-teeth (the name in Pliny refers to its use for treating toothache)

*odonto-*  tooth-

*odoratus -a -um, odorifer -era -erum, odorus -a -um*  fragrant, scented

*oedo-*  swelling-, becoming swollen-

*Oenanthe*  Wine-fragrant

*Oenothera*  Ass-catcher (the Greek name for another plant)

*officinalis -is -e, officinarum*  sold in shops, of shops, of the apothecaries, officinal medicines

*Oftia*  a name by Adanson with no clear meaning

*-oides, -oideus -a -um*  -like, -resembling

144

*Olax*   Furrow (the appearance given by the two-ranked leaves)

*olbia, olbios*   rich or from Hyères (Olbia), France

*Olea*   Oil (the ancient name for the olive)

*oleaginosus -a -um*   fleshy, rich in oil

*oleander*   from the Italian (for the olive-like foliage)

*Olearia*   Olive-like (the leaves of some species)

*olei-*   olive-, *Olea-*

*oleifer -era -erum*   oil-bearing

*olens*   fragrant, musty, stinking

*-olentus -a -um*   -fullness of, -abundance

*oleraceus -a -um*   of cultivation, vegetable, suitable for food, aromatic

*olidus -a -um*   stinking, smelling

*oligo-*   small-, feeble-, few-

*olitorius -a -um*   of vegetable gardens or gardeners, salad vegetable

*olusatrum*   Pliny's name for a black-seeded pot-herb

*olympicus -a -um*   from Mt Olympus, Greece, Olympian

*omeiensis -is -e*   from Mt Omei, China (Szechwan)

*omorika*   from the Serbian name

*omphalo-*   navel-

*Omphalodes*   Navelled (the fruit shape)

*onc-, onco-*   tumour-, hook-

*Oncidium*   Tumour (the warted crest of the lip)

*onegensis -is -e*   from Onega, USSR

*onites*   a name used by Dioscorides (of an ass or donkey)

*Onobrychis*   Ass-bray (a name in Pliny for a legume eaten greedily by asses)

*Onoclea*   Closed-cup (the sori are concealed by the rolled fronds)

*Ononis*   the classical name used by Dioscorides

*Onopordum(on)*   Ass-fart (its effect on donkeys)

*Onosma*   Ass-smell (said to attract asses)

*oo-*   egg-shaped-

*opacus -a -um*   dull, shady, not glossy or transparent

*opalus*   from the old Latin name for maple, *opulus*

*operculatus -a -um*   lidded, with a lid

*ophio-*   snake-like, snake-

*ophiocarpus -a -um*   with an elongate fruit, snake-like-fruited

*ophioglossifolius -a -um*   snake's-tongue-leaved

*Ophioglossum*   Snake-tongue (appearance of fertile part of frond – adder's tongue fern)

*Ophrys*   Eyebrow (the name in Pliny)

*opistho-*   back-, behind-

*oporinus -a -um*   of late summer, autumnal

*oppositi-*   opposite-, opposed-

*-ops, -opsis -is -e*   -like, -looking like, -appearance of

*opuli-*   guelder-rose-like-

*opulus*   an old generic name for the guelder rose

*Opuntia*   Tournefort's name for succulent plants from a Greek town of the same name

*opuntiiflorus -a -um (opuntiaeflorus -a -um)*   *Opuntia*-flowered

*orarius -a -um*   of the shoreline

*orbicularis -is -e, orbiculatus -a -um*   disc-shaped, circular in outline, orbicular

*orcadensis -is -e*   from the Orkney Isles, Orcadian

*orchioides*   resembling *Orchis*

*Orchis*   Testicle (the shape of the root-tubers)

*oreganus -a -um, oregonensis -is -e, oregonus -a -um*   from Oregon, USA

*orellana*   from a pre-Linnaean name for annatto

*oreo-, ores-, ori*   mountain-

*Oreodoxa*   Mountain-glory

*oreophilus -a -um*   mountain-loving, montane

*oresbius -a -um*   living on mountains

*organensis -is -e*   from Organ Mt, New Mexico, USA or Brazil

*orgyalis -is -e*   about 6 feet in length (the distance from finger-tip to finger-tip with arms stretched)

*orientalis -is -e*   eastern, oriental

*Origanum*   Theophrastus' name for an aromatic herb

*-orius -a -um*   -able, -capable of, -functioning

*ormo-*   necklace-like-, necklace-

*ornatus -a -um*   adorned, showy

*ornitho-*   bird-like-, bird-

*Ornithogalum*   Bird-milk (yields a bird-lime)

*ornithopodus -a -um*   bird-footed, like a bird's foot

*Ornithopus*   Bird-foot (the fruits)

*ornithorhynchus -a -um*   like a bird's beak

*ornus*   from the ancient Latin for manna-ash

*Orobanche*   Legume-strangler (one species is a parasite on legumes – see also *rapum-genistae*)

*orobus*   an old generic name for a leguminous plant

*orontium*   an old generic name for a plant from the Orontium river, Syria

*orophilus -a -um*   mountain-loving, montane

*orospendanus -a -um*   of mountains

146

*orphanidium*   fatherless, unrelated
*Orthila*   Straight (the style)
*ortho-*   correct-, upright-, straight-
*Orthocarpus*   Upright-fruit
*orubicus -a -um*   from Oruba Island, Caribbean
*orvala*   origin obscure, possibly from Greek for a sage-
  (horminon-)like plant
*Oryza*   from the Arabic 'eruz'
*Oryzopsis*   Oryza-resembler
*oscillatorius -a -um*   able to move about a central attachment,
  versatile
*-osma, osmo-*   -scented, fragrant-
*Osmanthus*   Fragrant-flower
*Osmunda*   either for Osmund the waterman or for the Anglo-
  Saxon equivalent of Thor, god of thunder
*osseus -a -um*   of very hard texture, bony
*ossifragus -a -um*   of broken bones (said to cause fractures in
  cattle when abundant in pastures)
*osteo-*   bone-like-, bone-
*Osteospermum*   Bone-seed (the hard-coated fruits)
*ostruthius -a -um*   purplish
*Ostrya*   a name in Pliny for a hornbeam
*-osus -a -um*   -abundant, -large, -very much
*-osyne, -otes*   -notably
*ot-, oto-*   ear-like-, ear-
*Otanthus*   Ear-flower (the shape of the corolla)
*Othonia*   Cloth-napkin (the covering of downy hairs)
*otites*   an old generic name, from Rupius, relating to ears
*Ottelia*   from the native Malabar name
*-otus -a -um*   -resembling, -having
*ouletrichus -a -um*   with curly hair
*Ouratea*   from the South American native name
*ovali-, ovalis -is -e*   egg-shaped in outline, oval
*ovati-, ovatus -a -um*   egg-shaped (in the solid or in outline) with
  the broad end lowermost
*ovifer -era -erum, oviger -era -erum*   bearing eggs (or egg-like
  structures)
*ovinus -a -um*   of sheep
*Oxalis*   Acid-salt (the name in Nicander refers to the taste)
*oxy-, -oxys*   acid-, sharp-, -pointed
*Oxyacantha*   Sharp-thorn (Theophrastus' name)
*Oxycedrus*   Pungent-juniper
*Oxycoccus*   Acid-berry

*oxygonus -a -um*   with sharp angles, sharp-angled
*oxylobus -a -um*   with sharp-pointed lobes
*oxyphilus -a -um*   of acidic soils, acid soil-loving
*Oxyria*   Acidic (the taste)
*Oxytropis*   Sharp-keel (the pointed keel petal)


*pabularis -is -e, pabularius -a -um*   of forage or pastures
*pachy-*   stout-, thick-
*pachyphloeus -a -um*   thick-barked
*Pachyphragma*   Stout-partition (the ribbed septum of the fruit)
*Pachysandra*   Thick-stamens (the filaments)
*pacificus -a -um*   of the western American seaboard
*padus*   Theophrastus' name for St Lucie Cherry
*Paeonia*   named by Theophrastus for Paeon, the physician who
   in mythology was changed into a flower by Pluto
*paganus -a -um*   of country areas, from the wild
*palaestinus -a -um*   from Palestine, Palestinian
*paleaceus -a -um*   covered with chaffy scales, chaffy
*palinuri*   from Palinuro, Italy
*Palisota*   for A. M. F. Palisot de Beauvois, French botanist
   (1752–1820)
*Paliurus*   the ancient Greek name for Christ-thorn
*pallens*   pale
*pallescens*   becoming pale, fading
*palliatus -a -um*   cloaked, hooded
*pallidus -a -um*   greenish, pale
*palmaris -is -e*   of a hand's breadth, about 3 inches wide
*palmati-, palmatus -a -um*   with five or more veins arising from
   one point (usually on divided leaves), palmate (see Fig. 5 (*a*))
*palmensis -is -e*   from Las Palmas, Canary Isles
*palmitifidus -a -um*   palmately incised
*paludosus -a -um, paluster -tris -tre*   of boggy or marshy ground
*pampinosus -a -um*   leafy
*Panax*   Healer-of-all (the ancient virtues of Ginseng)
*pancicii*   for Joseph Pančić, Yugoslavian botanist (1814–1888)
*Pancratium*   All-potent (a name used by Dioscorides)
*Pandanus*   Malayan name for screw-pines
*panduratus -a -um*   fiddle-shaped, pandurate
*paniceus -a -um*   like a millet grain
*paniculatus -a -um*   with a branched-racemose inflorescence,
   paniculate (see Fig. 2 (*c*))
*Panicum*   the ancient Latin name for *Setaria*
*pannonicus -a -um*   from south-west Hungary (Pannonia)

*pannosus -a -um*   woolly, tattered, coarse, ragged

*panormitanus -a -um*   from Palermo, Sicily

*Papaver*   the Latin name for poppies, including the opium poppy

*paphio-*   Venus'-

*papil-, papilio-*   butterfly-

*papilliger -era -erum, papillosus -a -um*   having papillae or minute lobes on the surface, papillate

*papulosus -a -um*   pimpled with small soft tubercles

*papyraceus -a -um*   with the texture of paper, papery

*papyrifer -era -erum*   paper-bearing

*Papyrus*   Paper (the Greek name for the paper made from the Egyptian bulrush, *Cyperus p.*

*para-*   near-, besides-

*parabolicus -a -um*   ovate-elliptic, parabolic in outline

*paradisi, paradisiacus -a -um*   of parks, of gardens, of paradise

*paradoxus -a -um*   strange, unusual, unexpected

*paraguariensis -is -e, paraguayensis -is -e*   from Paraguay

*paralias*   seaside (ancient Greek name for a plant)

*Parapholis*   Irregular-scales (the position of the glumes)

*parasiticus -a -um*   living on other plants, parasitic (it formerly included epiphytes)

*parci-*   with few-

*parcifrondiferus -a -um*   bearing few or small leafy shoots, with few-leaved fronds

*pardalianches, pardalianthes*   leopard-strangling (poisonous leopard's-bane)

*pardalinus -a -um, pardinus -a -um*   spotted or marked like a leopard

*pardanthinus -a -um*   resembling *Belamcanda* (*Pardanthus*)

*Parentucellia*   for Th. Parentucelli (Pope Nicholas V)

*pari-*   equal-, paired-

*Parietaria*   Wall-dweller

*parietarius -a -um*   of walls (a name in Pliny used for a plant growing on walls), parietal (the placentas on the wall of the ovary)

*Paris*   Equal (the regularity of its leaves and floral parts)

*parmulatus -a -um*   with a small round shield

*parnassi, parnassiacus -a -um*   from Mt Parnassus, Greece

*Parnassia*   Parnassus (the native home of Gramen Parnassi – grass of Parnassus)

*Paronychia*   Beside-nail (formerly used to treat whitlows)

*Parthenium*   a Greek name for composites with white ray florets

*Parthenocissus*   Virgin-ivy (Virginia creeper)

*parthenus -a -um*   virgin, virginal, pure, chaste

*-partitus -a -um*   -deeply divided, -partite, -parted

*parvi-, parvus -a -um*   small

*parvulus -a -um*   very small

*pascuus -a -um*   of pastures

*Paspalum*   a Greek name for millet grass

*Passiflora*   Passion-flower (the signature of the numbers of parts in the flower to the events of the Passion)

*Pastinaca*   food, eatable, from a trench in the ground (formerly for carrot and parsnip)

*pastoralis -is -e*   of shepherds, growing in pastures

*patagonicus -a -um*   from Patagonian area of South America

*patavinus -a -um*   from Padua, Italy

*patellaris -is -e, patelliformis -is -e*   knee-cap-shaped, small dish-shaped

*patens, patenti-*   spreading out from the stem, patent

*patientia*   patience (corruption of patience dock *Lapathum*)

*patulus -a -um*   spreading, opened up

*pauci-, paucus -a -um*   little-, few

*pauciflorus -a -um*   few-flowered

*Paulownia*   for Anna Pavlovna, daughter of Paul I of Russia (1795–1865)

*pauperculus -a -um*   poor

*Pavonia, pavonianus -a -um*   for Don José Pavón, Spanish botanist in Peru (1790–1844)

*pavonicus -a -um, pavoninus -a -um*   peacock-blue, peacock-like, showy

*pavonius -a -um*   peacock-blue, resembling *Pavonia*

*pecten-veneris*   Venus' comb (a name used in Pliny)

*pectinatus -a -um*   comb-like, pectinate

*pectinifer -era -erum*   with a finely divided crest, comb-bearing

*pectoralis -is -e*   of the chest (used to treat coughs)

*pedalis -is -e*   about a foot in length or stature

*pedatifidus -a -um, pedatus -a -um*   palmate but with the lateral divisions subdivided, pedate (see Fig. 5(*b*))

*pedemontanus -a -um*   from Piedmont, north Italy (foot of the hills)

*pedicellatus -a -um, pedicellaris -is -e*   each flower borne on its own stalk in the inflorescence

*Pedicularis*   Louse-wort

*pedicularis -is -e*   of lice (name of a plant in *Columella* thought to be associated with lice)

*pedifidus -a -um*   shaped like a (bird's) foot

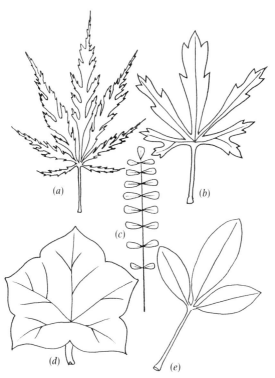

Fig. 5. Some leaf shapes which provide specific epithets:
(a) palmate (e.g. *Acer palmatum* Thunb. 'Dissectum'. As this maple's
leaves mature, the secondary division of the leaf-lobes passes
through incised-, *incisum*, to torn-, *laciniatum*, to dissected-,
*dissectum*, lobed, from one central point; (b) pedate (e.g. *Callirhoe
pedata* Gray). This is distinguished from palmate by having the
lower, side lobes themselves divided; (c) pinnate (e.g. *Ornithopus
pinnatus* Druce). When the lobes are more or less strictly paired it is
called paripinnate, when there is an odd terminal leaflet it is called
imparipinnate and when the lobing does not extend to the central
leaf-stalk it is called pinnatifid; (d) peltate (e.g. *Pelargonium peltatum*
(L.) Ait.) has the leaf-stalk attached on the lower surface, not at the
edge; (e) ternate (e.g. *Choisya ternata* H.B.K.). In other ternate leaves
the three divisions may be further divided, ternately, palmately, or
pinnately.

*pedil-, pedilo-*   shoe-, slipper-

*Pedilanthus*   Shoe-flower (involucre of bird cactus)

*peduncularis -is -e, pedunculatus -a -um*   with the inflorescence suported on a distinct stalk, pedunculate

*pedunculosus -a -um*   with many or conspicuous peduncles

*Peganum*   Theophrastus' name for rue

*pekinensis -is -e*   from Pekin, China

*pel-*   through-

*Pelargonium*   Stork (Greek name compares the fruit shape of florists' geraniums with a stork's head)

*pelios-*   livid-

*Pellaea*   Dusky (the fronds of most)

*pellucidus -a -um*   through which light passes, transparent, clear, pellucid

*pelorius -a -um*   monstrous, peloric (as with radial forms of normally bilateral flowers)

*peltatus -a -um*   stalked from the surface (not the edge), peltate (see Fig. 5 (*d*))

*Peltophorum*   Shield-bearer (the shape of the stigma)

*peltophorus -a -um*   with flat scales, shield-bearing

*pelviformis -is -e*   shallowly cupped, shaped like a shallow bowl

*pendens, penduli-, pendulinus -a -um, pendulus -a -um*   drooping, hanging down

*penicillatus -a -um, penicillius -a -um*   covered with tufts of hair, brush-like

*peninsularis -is -e*   living on a peninsula

*pennatus -a -um, penni-, penniger -era -erum*   arranged like the barbs of a feather, feathered

*penninervis -is -e*   pinnately nerved

*Pennisetum*   Feathery-bristle

*pennsylvanicus -a -um, pensylvanicus -a -um*   from Pennsylvania, USA

*pensilis -is -e*   hanging down, pensile

*pen, pent-, penta-*   five-

*Penstemon (Pentstemon)*   Five-stamens

*pentaglottis*   Five-tongues (the scales in the throat of the corolla)

*Peperomia*   Pepper-like (some resemble *Piper*)

*peplis*   Dioscorides' name for a Mediterranean coastal spurge

*peploides*   peplus-like

*peplus*   Dioscorides' name for a northern equivalent of *peplis*

*Pepo*   Sun-cooked (ripens to become edible)

*per-*   through-, beyond-, extra-, very-

*peramoenus -a -um*   very beautiful, very pleasing

*percussus -a -um*   actually or appearing to be perforated
*peregrinus -a -um*   strange, foreign, exotic
*perennans, perennis -is -e*   through the years, continuing, perennial
*Pereskia*   for N. C. F. de Pieresc (1580–1637)
*perfoliatus -a -um, perfossus -a -um*   the stem appearing to pass through the completely embracing leaves
*perforatus -a -um*   pierced or apparently pierced with round holes
*Pergularia*   Arbour (the twining growth)
*peri-*   around-
*periclymenum*   Dioscorides' name for a twining plant
*Periploca*   Twine-around
*permixtus -a -um*   confusing
*Pernettya*   for A. J. Pernetty, traveller and writer (1716–1801)
*peronatus -a -um*   with woolly-mealy covering (on fungal fruiting bodies)
*perpropinquus -a -um*   very closely related
*perpusillus -a -um*   exceptionally small, very small
*Persicaria*   Peach-like (the leaves)
*persicarius -a -um*   resembling peach (the leaves), an old name for *Polygonum hydropiper*
*persici-*   peach-
*persicus -a -um*   from Persia, Persian
*persistens*   persistent
*personatus -a -um*   with a two-lipped mouth
*perspicuus -a -um*   transparent
*persutus -a -um*   with slits or holes
*pertusus -a -um*   pierced through, perforated
*perulatus -a -um*   wallet-like, with conspicuous scales (e.g. on buds)
*peruvianus -a -um*   from Peru, Peruvian
*perviridis -is -e*   deep-green
*-pes*   -stalk, -foot
*pes-caprae*   goat's foot
*pes-tigridis*   tiger's foot
*Petasites*   Sun-hat (Dioscorides' name refers to the large leaves)
*petecticalis -is -e*   blemished with spots
*petiolaris -is -e, petiolatus -a -um*   having a petiole, not sessile, petiolate
*petiolosus -a -um*   with conspicuous petioles
*petraeus -a -um, petr-, petro-*   stony, rocky, of rocky places
*Petrophila*   Rock-lover (habitat preference)

*Petroselinum*   Dioscorides' name for parsley
*Petunia*   from the Brazilian name for tobacco 'petun'
*Peucedanum*   a name used by Theophrastus for hog fennel
*Peyrousea*   see *Lapeirousia*
*phaeno-*   shining-
*phaeo-, phaeus -a -um*   dark-, dusky-brown
*Phalaenopsis*   Moth-like (flowers of the moth orchid)
*Phalaris*   Helmet-ridge (Dioscorides' name for a plume-like grass)
*phanero-*   conspicuous-
*Phaseolus*   the old Latin name for a kind of bean
*Phegopteris*   Oak-fern
*Phelipaea (Phelypaea)*   for Louis Phelipeaux, patron of J. P. de Tournefort
*Phellandrium (Phellandrion)*   a name in Pliny for an ivy-leaved plant
*phello-*   corky-, cork-
*Phellodendron*   Cork-tree (the thick bark)
*phil-, philo-, -philus -a -um*   loving-, liking-, -fond of
*philadelphicus -a -um*   from Philadelphia
*Philadelphus*   Brotherly-love
*philippensis -is -e*   from the Philippines
*Philonotis*   Moisture-lover
*phlebanthus -a -um*   with veined flowers
*phleioides*   rust-like, resembling *Phleum*
*Phleum*   Copious (Greek name for a kind of dense-headed rush)
*-phloebius -a -um*   -veined
*-phleus -a -um*   -barked, -bark
*phlogi-*   flame-, *Phlox*-like
*Phlomis*   Flame (the hairy leaves were used as lamp wicks)
*Phlox*   Theophrastus' name for a plant with flame-coloured flowers
*phoeniceus -a -um*   red-purple, from Tyre and Sidon (Phoenicia)
*phoenicolasius -a -um*   purple-haired
*Phoenix*   Phoenician (who introduced the date palm to the Greeks)
*Pholiurus*   Scale-tail
*Phormium*   Basket (from the use of the leaf-fibres)
*-phorus -a -um*   -bearing, -carrying
*Phragmites*   Hedge-dweller, (common habitat)
*phrygius -a -um*   from Phrygia, Asia Minor
*phu*   foul-smelling
*Phuopsis*   valerian-like, resembling *Valeriana phu*

*Phylica*   Leafy (the copious foliage)
*phylicifolius -a -um*   with leaves like those of *Phylica*
*phyll-*   leaf-
*Phyllanthus*   Leaf-flower (some flower from leaf-life phyllodes)
*Phyllitis*   Dioscorides' name refers to the simple leaf-like frond
*Phyllodoce*   the name of a sea nymph
*phyllomaniacus -a -um*   excessively leafy, a riot of foliage
*-phyllus -a -um*   -leaved
*physa-*   bladder-
*Physalis*   Bellows (the inflated fruiting calyx)
*physo-*   bellows-, inflated-
*Physocarpus*   Inflated-fruit
*physodes*   puffed out, inflated-looking
*Physospermum*   Inflated-seed (fruit of bladder seed)
*Phyteuma*   a name used by Dioscorides for rampion
*phyto*   plant-
*Phytolacca*   Plant-dye (the sap of the fruit)
*Picea*   Pitch (the resin)
*piceus -a -um*   black, blackening
*picridis -is -e*   ox-tongue-like, of *Picris*
*Picris*   Bitter (Theophrastus' name for a bitter potherb)
*picturatus -a -um*   variegated
*pictus -a -um*   brightly marked, painted
*pileatus -a -um*   capped, having a cap
*piliferus, pilifer -era -erum*   bearing hairs, with short soft hairs,
   ending in a long fine hair
*pilo-, pilosus -a -um*   felted with long soft hairs, pilose
*pilosellus -a -um*   tomentose, finely felted with soft hairs
*Pilularia*   Small-balls (the sporocarps)
*pilularis -is -e, pilulifer -era -erum*   bearing small balls or globular
   structures
*Pimpinella*   a medieval name of uncertain meaning
*pimpinellifolius -a -um*   *Pimpinella*-leaved
*pinaster*   Pliny's name for *Pinus sylvestris*
*pindicola, pindicus -a -um*   from the Pindus mountains of north
   Greece
*pinetorum*   of pine woods
*pineus -a -um*   cone-producing, of pines, resembling a pine
*pingui-*   fat-
*Pinguicula*   Fat (the leaves)
*pini-*   pine-like, pine-
*pinnati-, pinnatus -a -um*   set in two opposite ranks, winged,
   feathered, pinnate (see Fig. 5(*c*))

*pinnatifidus -a -um*   pinnately divided

*pinsapo*   from the Spanish name 'pinapares'

*Pinus*   the Latin name for a pine

*Piper*   from the Indian name for pepper

*piperascens*   pepper-like, resembling *Piper*

*piperitus -a -um*   peppermint scented, peppery (the taste)

*piperinus -a -um*   peppery (-scented)

*Piptanthus*   Falling-flower (quickly deciduous bloom)

*piri-*   pear-

*piriformis -is -e*   pear-shaped

*Pirola*   Small-pear (similarity of foliage)

*Pirus*   the Latin name for a pear tree

*pisi-, piso-*   pea-like-, pea-

*pisifer -ear -erum*   bearing peas

*pissardii (pissardi, pissarti)*   for M. Pissard who introduced
 *Prunus, cerasifera* 'Pissardii'

*Pistacia*   the Greek name used by Nicander in 200 BC

*Pistia*   Watery (habitat of the water lettuce) and derived from
 the Persian 'foustag'

*Pisum*   the Latin name for the pea

*pithece-, pitheco-*   ape-, monkey-

*Pithecellobium*   Monkey-ears (the shape of the fruit)

*Pittosporum*   Pitch-seed (the resinous coating of the seed)

*placatus -a -um*   quiet, calm, gentle

*plagio-*   oblique-

*planeta, planetes*   not stationary, planet-like, wandering

*plani-, planus -a -um*   flat

*planiusculus -a -um*   flattish, somewhat flat

*plantagineus -a -um*   ribwort-like, plantain-like

*Plantago*   Foot-sole (the way the leaves of some lie flat on the
 ground)

*plat-, platy-*   broad-, flat-

*Platanthera*   Flat-anthers

*Platanus*   Flat-leaf (the Greek name for a plane tree)

*platycentrus -a -um*   wide-eyed, broad-centred

*Platycodon*   Wide-bell (the flower form)

*plebio-, plebius -a -um*   common

*plecto-, plectus -a -um*   pleated

*plectro-, plectrus -a -um*   spur-, spurred

*pleio-, pleo-*   many-, more-, full-, large-, thick-, several-

*pleiospermus -a -um*   thick-seeded

*pleni-, plenus -a -um*   double, full

*pleniflorus -a -um*   double-flowered

*plenissimus -a -um*   very full or double flowered

*plesio-*   near to-, close by-

*pleuri-, pleuro-*   ribs, edge-, side-, of the veins-

*plicati- plicatus -a -um, plici-, ploco-*   pleated, folded lengthwise, plicate

*plumarius -a -um, plumatus -a -um*   feathery, plumed, plumose

*Plumbago*   Leaden (Pliny's name refers to the flower colour)

*plumbeus -a -um*   lead-coloured

*plumosus -a -um*   feathery

*pluri-*   many-, several-

*pluriflorus -a -um*   many-flowered

*pluvialis -is -e, pluviatilis -is -e*   announcing rain, growing in rainy places

*pneumonanthe*   lung-flower (the former use of *Gentiana p.* for respiratory disorders)

*Poa*   Pasturage (the Greek name for a grass)

*pocophorus -a -um*   fleece-bearing

*poculiformis -is -e*   cup-shaped with upright limbs (corollas)

*podagrarius -a -um, podagricus -a -um*   snare, of gout (used to treat gout)

*Podalyra*   for Podalyrius, son of Aesculapius

*podalyriaefolius -a -um*   with leaves resembling those of *Podalyra*

*-podioides*   -foot-like

*-podius -a -um, podo-, -podus -a -um*   foot, stalk

*Podocarpus*   Foot-fruit (the characteristic shape)

*poecilo-*   variable-, variegated-, spotted-

*poetarum, poeticus -a -um*   of poets (Greek gardens included games areas and theatres)

*-pogon*   -haired, -bearded

*poikilo-*   variable-, variegated-, spotted-

*poissonii*   for M. Poisson, French botanist

*polaris -is -e*   from the North polar region, of the North Pole

*Polemonium*   for King Polemon of Pontus (the name used by Pliny)

*polifolius -a -um*   grey-leaved, *Teucrium*-leaved

*polio-*   grey-

*politus -a -um*   elegant, polished

*polium*   greyish-white (foliage)

*pollicaris -is -e*   as long as the end joint of the thumb, about one inch

*polonicus -a -um*   from Poland, Polish

*poly-*   many-

*polyacanthus -a -um*   many-spined

*polyanthemos, polyanthus -a -um*   many flowered

*Polycarpon*   Many-fruits (named by Hippocrates)

*polyedrus -a -um*   many-sided

*Polygala*   Much-milk (Dioscorides' name refers to the improved lactation in cattle fed on it)

*polygamus -a -um*   the flowers having various combinations of the reproductive structures

*Polygonatum*   Many-knees (the structure of the rhizome)

*Polygonum*   Many-joints (the swollen stem nodes)

*polygyrus -a -um*   twining

*polymorphus -a -um*   variable, of many forms

*polypodioides*   resembling *Polypodium*

*Polypodium*   Many-feet (the rhizome growth)

*Polypogon*   Many-bearded

*Polystichum*   Many-rows (the arrangement of the sori on the fronds)

*pomeridianus -a -um*   of the afternoon (afternoon flowering)

*pomi-, pomaceus -a -um*   apple-like

*pomifer -era -erum*   apple-bearing, pome-bearing

*pomponius -a -um*   of great splendour, pompous, having a top-knot or pompon

*ponderosus -a -um*   heavy, large, ponderous

*Pontederia*   for Guillo Pontedera, former Professor of Botany at Padua (1688–1757)

*ponticus -a -um*   of the Black Seas's southern area, Pontus

*populifolius -a -um*   poplar-leaved

*populneus -a -um*   poplar-like, related to *Populus*

*Populus*   the ancient name for poplar, *arbor populi* 'tree of the people'

*porcinus -a -um*   of pigs

*porophilus -a -um*   loving soft stony ground

*porophyllus -a -um*   having holes in the leaves

*porosus -a -um*   with holes or pores

*porphyreus -a -um, porphyrion*   warm-reddish-coloured

*porri-*   leek-like-, leek-, *porrum*-like-

*porrifolius -a -um*   leek-leaved

*porrigens*   spreading

*porrigentiformis -is -e*   *porrigens*-like (with leaf-margin teeth pointed outwards and forwards)

*porrum*   a Latin name used for various *Allium* species

*portensis -is -e*   from Oporto

*portlandicus -a -um*   from the Portland area

*portoricensis -is -e*   from Puerto Rico
*portula*   abbreviated form of *Portulaca*
*Portulaca*   Milk-carrier (a name in Pliny)
*portulaceus -a -um*   *Portulaca*-like
*post-*   behind-, after-, later-
*posticus -a -um*   turned outwards from the axis, extrorse
*potam-*, *potamo-*   watercourse-, of watercourses-
*Potamogeton*   Watercourse-neighbour (the habitat)
*potamophilus -a -um*   river-loving
*potatorum*   of drinkers (used for fermentation)
*Potentilla*   Quite-powerful (as a medicinal herb)
*Poterium*   Drinking-cup (a name used previously for another plant)
*poukhanensis -is -e*   from Pouk Han, Korea
*-pous*   -foot, -stalk, -stalked
*prae-*   before-, in front-
*praealtus -a -um*   very tall or high
*praecox*   earlier than most of its genus, forward, very early developing
*praegnans*   full, swollen, pregnant(-looking)
*praemorsus -a -um*   as if nibbled at the tip
*praeruptorum*   of rough places (living on screes)
*praestans*   excelling, distinguished
*praetermissus -a -um*   overlooked, omitted
*praetextus -a -um*   bordered
*praevernus -a -um*   before spring, early, prevernal
*prasinus -a -um*, *prasus -a -um*   leek-green, leed-like, from a Latin name for various species of *Allium*
*pratensis -is -e*   of meadows
*pratericolus -a -um*, *praticolus -a -um*   of meadows, living in grassy places
*pravissimus -a -um*   very crooked
*precatorius -a -um*   relating to prayer (rosary beads)
*prenans*   drooping
*Prenanthes*   Drooping flower (the nodding flowers)
*preptus -a -um*   eminent
*Primula*   Little-firstling (Spring flowering)
*primulinus -a -um*   primrose-coloured, *Primula*-like
*primuloides*   resembling *Primula*
*princeps*   most distinguished, first, princely
*priono-*   serrated-, saw-toothed-
*Prionum*   Saw (the leaf-margins)
*prismati-*, *prismaticus -a -um*   prism-, prism-like-

159

*pro-* forwards-, before-, for-, instead of-
*proboscoides, proboscoideus -a -um* snout-like, trunk-like
*procerus -a -um* very tall
*procumbens* lying flat on the ground, creeping forwards, procumbent
*procurrens* spreading below ground, running forwards
*prodigiosus -a -um* wonderful, marvellous, prodigious
*productus-a -um* stretched out, extended, produced
*profusus -a -um* very abundant, profuse
*prolifer -era -erum* producing offsets or young plantlets or bunched growth, proliferous
*prolificus -a -um* very beautiful, very fruitful
*pronus -a -um* lying flat, with a forward tilt
*propaguliferus -a -um* prolific, multiplying by vegetative propagules
*propensus -a -um* hanging down
*propinquus -a -um* closely allied, of near relationship, related
*pros-* near-, in addition-, also-
*proso-, prostho-* towards-, to the front-, before-
*prostratus -a -um* lying flat but not rooting, prostrate
*Protea* for Proteus (the sea god's versatility in changing form)
*protrusus -a -um* protruding
*provincialis -is -e* from Provence, France
*pruhonicus -a -um* from Průhonice, Czechoslovakia
*pruinatus -a -um, pruinosus -a -um* with a glistening surface as though frosted over, with hoary bloom
*Prunella* (*Brunella*) from the German name for quinsy 'die Bräune' for which it was used as a cure
*pruni-* plum-like, plum-
*Prunus* the Latin name for a plum tree
*pruriens* irritant, stinging
*Psamma* Strand-dweller (an old generic name for marram grass refers to its habitat)
*pseud-, pseudo-* sham-, false-
*Pseuderanthemum* False-*Eranthemum*
*Pseudotsuga* False-*Tsuga*
*Psidium* a Greek name *psidion* formerly for the pomegranate (similarity of the fruits)
*psilo-* smooth-, bare-
*psilostemon* slender- or naked-stamened
*Psilotum* Hairless
*psittacinus -a -um* parrot-like (the coloration)
*psittacorum* of parrots

*Psoralea*   Manged (the dot-marked vegetative parts)
*psycodes*   fragrant, butterfly-like
*psyllium*   of fleas (from a Greek name, referring to the appearance of the seeds)
*ptarmicoides*   ptarmica-like, resembling *Achillea ptarmica*
*ptarmicus -a -um*   causing sneezes
*ptera-, ptero-, -pteris -is -e, -pterus -a -um*   -winged
*pteranthus -a -um*   with winged flowers
*Pteridium*   Small-fern
*Pteris*   Feathery (the Greek name for a fern)
*Pterocarya*   Wing-nut (the winged fruits of most)
*ptilo-*   feathery-
*ptycho-*   folded-
*pubens, pubescens, pubiger -era -erum*   softly hairy, covered with down, downy, pubescent
*Pubilaria*   Hairy (the clothing of fibrous leaf remains on the rhizome)
*Puccinellia*   for B. Puccinelli, Italian botanist of Lucca (1808–1850)
*puddum*   from a Hindi name for a cherry
*puderosus -a -um*   very bashful
*pudicus -a -um*   retiring, modest, bashful
*puellii*   fof Timothée Puel, French botanist of Paris (1812–1890)
*pugioniformis -is -e*   dagger-shaped
*pulchellus -a -um*   beautiful, pretty
*pulcher -ra -rum*   beautiful, handsome, fair
*pulcherrimus -a -um*   most beautiful, most handsome
*Pulegium*   Flea-dispeller (a Latin plant name)
*Pulicaria*   Latin name for a plant which wards off fleas (Fleabane)
*pulicaris -is -e*   of fleas (e.g. the shape of the fruits)
*pullatus -a -um*   clothed in black, sad-looking
*pullus -a -um*   raven-black, almost dead-black
*Pulmonaria*   Lung-wort (the signature of the spotted leaves as indicative of efficacy in the treatment of respiratory disorders)
*pulposus -a -um*   fleshy
*Pulsatilla*   Quiverer (the pulsating movement of the flowers in the wind)
*pulverulentus -a -um*   covered with powder, powdery
*pulvinatus -a -um*   cushion-like, cushion-shaped
*pumilus -a -um*   low, small, dwarf
*punctati-, puncti-, punctatus -a -um*   with a pock-marked surface, spotted, punctate

161

*puncticulatus -a -um*  minutely spotted
*punctilobulus -a -um*  dotted-lobed
*pungens*  ending in a sharp point, pricking
*Punica*  Carthaginian (from a name in Pliny, *malum punicum*)
*puniceus -a -um*  crimson, carmine-red
*purgans, purgus -a -um*  purgative
*purpurascens*  becoming purple
*purpuratus -a -um*  empurpled, purplish
*purpureus -a -um*  reddish-purple
*purpusii*  for either of the brothers J. A. and C. A. Purpus of
  Darmstadt
*-pus*  -foot
*pusillus -a -um*  insignificant, minute, very small
*pustulatus -a -um*  covered with blisters, pustules
*putens*  foetid, stinking
*puteorum*  of the pits
*pycn-, pycno-*  compact-, densely-, dense-
*pycnanthus -a -um*  densely-flowered
*pycnostachyus -a -um*  thick-spiked
*pygmaeus -a -um, pygmeus -a -um*  dwarf
*Pyracantha*  Fire-thorn (persistent irritation from the thorn
  pricks, and the appearance in fruit)
*Pyramidalis -is -e, pyramidatus -a -um*  conical, pyramidal
*pyrenaeus -a -um, pyrenaicus -a -um*  from the Pyrenees
*Pyrethrum*  Fire (medicinal use in treating fevers)
*pyri-*  pear-
*pyriformis -is -e*  pear-shaped
*Pyrola*  Pear-like (compares the leaves with those of *Pyrus*)
*Pyrus*  the ancient Latin name for a pear tree
*pyxidatus -a -um*  with a lid, box-like (e.g. some stamens)

*quadrangularis -is -e, quadrangulatus -a -um*  with four angles,
  quadrangular
*quadratus - -um*  four-sided, square-stemmed
*quadri-*  four-
*quadriauritus -a -um*  four-lobed, four-eared
*quadrifidus -a -um*  divided into four, cut into four
*Quamoclit*  from the Greek name formerly for a bean
*quaternarius -a -um, quaternellus -a -um*  with four divisions,
  four-partite
*querci-, quercinus -a -um*  oak-, oak-like, resembling *Quercus*
*Quercus*  the old Latin name for an oak
*-quetrus -a -um*  -angled, -acutely-angled

*quichiotis*   chimerical, quixotic

*quin-*, *quinque-*   five-

*quinatus -a -um*   five-partite, with five divisions, in fives

*quinquelocularis -is -e*   five-celled, five-locular (the ovary)

*quinquevulnerus -a -um*   with five marks (e.g. on the corolla), five-wounded

*Quisqualis*   Who? What? (from a Malay name 'udani' which Rumphius transliterated as Dutch, 'hoedanig' meaning How? What?

*quitensis -is -e*   from Quito, Ecuador


*racemi-*, *racemosus -a -um*   with flowers arranged in a raceme (see Fig. 2(*b*))

*radens*   scraping (the rough surface)

*radians*, *radiatus -a -um*   radiating outwards

*radicans*   with rooting stems

*radicatus -a -um*, *radicosus -a -um*   with a large, conspicuous or numerous roots

*radiiflorus -a -um*   with radiating flowers

*Radiola*   Radiating (the branches)

*radiosus -a -um*   having many rays

*radula*   rough, rasping, like a rasp

*ragusinus -a -um*   from Dubrovnik (Ragusa), Yugoslavia

*ramentaceus -a -um*   covered with scales (ramenta)

*-rameus -a -um*   -branched

*rami-*   branches-, of branches-, branching-

*Ramischia*   for F. X. Ramisch, Bohemian botanist (1798–1859)

*ramosissimus -a -um*   greatly branched

*ramosus -a -um*   branched

*ramulosus -a -um*   twiggy

*Ranunculus*   Little-frog (the amphibious habit of many)

*rapa*   an old Latin name for a turnip

*rapaceus -a -um*   of turnips, *Rapa*-like

*raphani-*   radish-, radish-like-

*Raphanus*   the Latin name for a radish

*Raphia*   Needle (the sharply pointed fruit)

*Rapistrum*   Rape-flower (implies inferiority of wild mustard)

*rapum-genistae*   rape-of-broom (parasite of *Sarothamnus*)

*rapunculoides*   resembling *Rapunculus*, rampion-like

*Rapunculus*   Little-turnip (the swollen roots)

*rari-*, *rarus -a -um*   uncommon, scattered

*rariflorus -a -um*   having scattered flowers

*Ravenala*   from the Madagascan name for the travellers' tree *ravus -a -um*   tawny-grey-coloured

*re-*   back-, again-, against-

*reclinatus -a -um*   drooping to the ground, reflexed, reclined

*rectus -a -um*   straight, upright, erect

*recurvatus -a -um, recurvus -a -um*   curved backwards

*recutitus -a -um*   circumcised (the appearance of the flower heads with the rays reflexed)

*redivivus -a -um*   coming back to life, renewed (perennial habit or reviving after drought)

*reductus -a -um*   drawn back, reduced

*reflexus -a -um*   bent back upon itself, reflexed

*refractus -a -um*   abruptly bent, splitting open

*regalis -is -e*   outstanding, kingly, royal, regal

*regerminans*   regenerating

*regina, reginae*   queen, of the queen

*regis-jubae*   King-Juba, who was a king of Numibia

*regius -a -um*   splendid, royal, kingly

*rehderi*   either for Jacob Heinrich Rehder of Moscow or for Alfred Rehder of the Arnold Arboretum, USA

*religiosus -a -um*   sacred, venerated, of religious rites (Buddha is reputed to have received enlightenment beneath the bo fig tree)

*remotus -a -um*   scattered (e.g. the flowers on the stalk)

*reniformis -is -e*   kidney-shaped, reniform

*repandens, repandus -a -um*   with a slightly wavy margin, repand

*repens*   creeping (stoloniferous)

*replicatus -a -um*   double-pleated, doubled down

*reptans*   creeping

*reptatrix*   creeping-rooted

*Reseda*   Healer (the name in Pliny refers to its use in treating bruises)

*resinifer -era -erum, resinosus -a -um*   producing resin, resinous

*resupinatus -a -um*   inverted (e.g. those orchids with twisted ovaries), resupine

*reticulatus -a -um*   netted, conspicuously net-veined, reticulate

*retortus -a -um*   twisted back

*retro-, retroflexus -a -um, retrofractus -a -um, retrorsus -a -um*   directed backwards and downwards

*retusus -a -um*   shallowly notched at the tip (e.g. leaves, see Fig. 7(*f*)), retuse

*revolutus -a -um*   rolled back, rolled out and under (e.g., leaf margin), revolute

*rex*   king

*rhabdotus -a -um*  striped

*rhaeticus -a -um*  from the Rhaetian Alps of the Swiss-Austrian border

*-rhagius -a -um*  -torn, -rent

*Rhamnus*  an ancient name for various prickly shrubs

*rhaponticus -a -um*  from the Black Sea area

*Rheum*  from a Parisian name for rhubarb

*Rhinanthus*  Nose-flower

*rhiz-, rhizo-, -rhizus -a -um*  root-, -rooted

*Rhizophora*  Root-carrier (the long-arched prop-roots)

*rhizophyllus -a -um*  root-leaved (the leaves form roots)

*rhod-, rhodo-*  rose-, rosy-, red-

*rhodantherus -a -um*  with red stamens

*rhodanthus -a -um*  rose-flowered

*rhodensis -is -e, rhodius -a -um*  from the Aegean Island of Rhodes

*Rhododendron(um)*  Rose-tree (a name formerly used for an oleander)

*rhodopaeus -a -um, rhodopensis -is -e*  from Rhodope Mt, Bulgaria

*Rhodotypos*  Rose-pattern (floral resemblance)

*rhoeas*  the old generic name of the field poppy

*Rhoicissus*  Pomegranate-ivy

*rhombi-, rhombicus -a -um, rhomboidalis -is -e, rhomboidosus -a -um*  diamond-shaped, rhombic

*rhopalo-*  club-, cudgel-

*rhumicus*  from the River Rhume area, West Germany

*Rhus*  from an ancient Greek name for a sumach

*Rhynchelytrum(on)*  Beaked-sheath (the shape of the glumes)

*Rhynchosia*  Beak (the shape of the keel petals)

*Rhynchosinapis*  Beaked-*Sinapis*

*Rhynchospora*  Beaked-seed

*rhytido-*  wrinkled-

*rhytidophyllus -a -um*  with wrinkled leaves

*Ribes*  from the Persian for acid-tasting

*Ricinus*  Tick (the appearance of the seeds)

*rigens, rigidus -a -um*  stiff, rigid

*rigensis -is -e*  from Riga, on the Baltic

*rimosus -a -um*  with a cracked surface, furrowed

*ringens*  with a two-lipped mouth, gaping

*riparius -a -um*  of the banks of streams and rivers

*ritro*  a southern European name for *Echinops ritro*

*rivalis -is -e*  of brooksides and streamsides

*Rivina, riviniana*   for A. Q. Rivinus, formerly Professor of Botany at Leipzig (1652–1722)

*rivularis -is -e*   of the waterside, of streamsides

*robbiae*   for Mary Anne Robb, who introduced *Euphorbia amygdaloides* ssp. *robbiae* from Turkey (1829–1912)

*robertianus -a -um*   of Robert (which Robert is uncertain)

*Robinia*   for Jean Robin, herbalist to Henry VI of France (1550–1629)

*robur*   oak timber, strong, hard

*robustus -a -um*   strong-growing, robust

*Roemeria*   for J. J. Römer, Swiss botanist (1763–1819)

*romanus -a -um*   of Rome, Roman

*Romulea*   for Romulus, founder of Rome

*roribaccus -a -um*   dewberry

*roridus -a -um*   with apparently minute blisters all over the surface, dewy

*Rorippa*   from an old Saxon name

*rorulentus -a -um*   dewy

*Rosa*   the Latin name for various roses

*rosaceus a -um*   looking or coloured like a rose

*rosae-, rosi-, roseus -a -um*   rose-like, rose-coloured

*Rosmarinus*   Seaside-dew (an ancient Latin name)

*rostellatus -a -um*   with a small beak, beaked

*rostratus -a -um*   with a long straight hard point, beaked, rostrate

*Rosularia*   Little-rose (the leaf rosettes)

*rosularis -is -e, rosulatus -a -um*   with leaf rosettes

*rotatus -a -um*   flat and circular, wheel-shaped

*rotundi-, rotundus -a -um*   rounded in outline or at the apex, spherical

*Roystonia*   for Gen. Roy Stone, American soldier (1836–1905)

*-rrhizus -a -um*   -rooted

*rubellinus -a -um, rubellus -a -um*   reddish

*rubens*   blushed with red, ruddy

*ruber, rubra, rubrum, rubri-, rubro-*   red

*rubescens, rubidus -a -um*   turning red, reddening

*Rubia*   a name in Pliny for madder

*rubicundus -a -um*   ruddy, reddened

*rubiginosus -a -um, rubrus -a -um*   rusty-red

*Rubus*   the Latin name for brambles

*Rudbeckia*   for Linnaeus' mentor Olof O. Rudbeck (1660–1740) and his son J. O. Rudbeck (1711–1790)

*ruderalis -is -e*   of waste places, of rubbish tips

*rudis -is -e*   untilled, rough, wild

*rudiusculus -a -um*   wildish

*rufescens, rufidus -a -um*   reddish, turning red

*rufinus -a -um*   red

*rufus -a -um*   rusty (-haired), pale- or reddish-brown

*rugosus -a -um*   wrinkled, rugose (e.g. leaf surfaces)

*rugulosus -a um*   somewhat wrinkled

*Rumex*   a name in Pliny for sorrel

*rumici-*   dock-like-

*ruminatus -a -um*   thoroughly mingled

*runcinatus -a -um*   sharply cut (leaf margins), saw-toothed with
   the fine tips pointing to the base

*rupester -tris -tre, rupicola*   of rock, of rocky places

*rupifragus -a -um*   growing in rock crevices, rock-cracking

*Ruppia (Ruppa)*   for H. B. Ruppius, German botanist
   (1688–1719)

*rupri-*   or rocks-, of rocky places-

*ruralis -is -e*   of country places, rural

*rurivagus -a -um*   of country roads, country wandering

*Ruschia*   for E. Rusch, South African farmer

*rusci-*   box-holly-like, butcher's-broom-like, resembling *Ruscus*

*Ruscus*   an old name for a prickly plant

*russatus -a -um*   reddened, russet

*russotinctus -a -um*   red-tinged

*rusticanus -a -um, rusticus -a -um*   of wild places, of the
   countryside, rustic

*Ruta*   the Latin name for rue

*ruta-baga*   from a Swedish name

*ruta-muraria*   rue-of-the-wall

*ruthenicus -a -um*   from Ruthenia, Russia

*rutilans, rutilus -a -um*   deep bright glowing red, orange-red

*rytido-*   wrinkled-


*sabaudus -a -um*   from Savoy (Sabaudia), south-east France

*sabatius -a -um*   from Capo di Noli, Riviera di Ponente, Italy

*sabbatius -a -um*   from Savona, north-west Italy

*sabdariffa*   from a West Indian name

*sabrinae*   from the River Severn (Sabrina)

*sabulicolus -a -um*   living in sandy places, sand-dweller

*sabulosus -a -um*   full of sand, of sandy ground

*saccatus -a -um*   bag-shaped, saccate

*saccharatus -a -um*   with a scattered white coating, sugared,
   sweet-tasting

*sacchariferus -a -um*   sugar-producing, bearing sugar

*saccharinus -a -um, saccharus -a -um*   sweet, sugary

*saccifer -era -erum*   having a hollowed part, bag-bearing

*sachalinensis -is -e*   from Sakhalin Island, eastern USSR

*sacrorum*   of sacred places, of temples, sacred (former ritual use)

*saepium*   of hedges

*Sagina*   Fodder (the virtue of a former included species, spurrey)

*sagittalis -is -e, sagittatus -a -um, sagitti*   arrow-shaped, sagittate
  (see Fig. 6(*c*))

*Sagittaria*   Arrowhead (the shape of the leaf-blades)

*sago*   yielding the large starch grains 'sago'

*Salaxis*   an unexplained name by Salisbury

*salebrosus -a -um*   rough

*salicarius -a -um, salice-, salici-, salicinus -a -um*   willow-like,
  willow-

*salicetorum*   of willow thickets

*Salicornia*   Salt-horn

*salignus -a -um*   of willow-like appearance, resembling *Salix*

*salinus -a -um*   of salt-marshes, halophytic

*salisburgensis -is -e*   from Salzburg, Austria

*Salix*   the Latin name for willows

*salpi-*   trumpet-

*Salpichroa*   Tube-of-skin (the flower)

*Salpiglossis*   Trumpet-tongue (the shape of the style)

*Salsola*   Salt (the taste and the habitat)

*salsuginosus -a -um*   of salt-marshes, living in saline soils

*salsus -a -um*   of saline soils, salted

*saltatorius -a -um*   dancing

*saltitans*   jumping

*saltuum*   of glades, woodlands or ravines

*salutaris -is -e*   healing, beneficial

*Salvia*   Healer (the old Latin name for sage)

*salvii-*   sage-like-, resembling *Salvia*

*Salvinia*   for A. M. Salvini, Italian botanist (1633–1722)

*salviodorus -a -um*   sage-scented

*saman, Samanea*   from a South American name

*sambucinus -a -um*   elder-like, resembling *Sambucus*

*Sambucus*   the Greek name for the elder tree

*samius -a -um*   from the Isle of Samos, Greece

*Samolus*   from a Celtic Druidic name

*sanctus -a -um*   holy

*sanguinalis -is -e, sanguineus -a -um, sanguineolentus -a -um*
  blood-red bloody

*Sanguisorba*   Blood-stauncher (has styptic property)

*Sanicula*   Little-healer (its medicinal property)

*Santolina*   Holy-flax

*Santalum*   from the Persian name 'shandal' for sandalwood

*Sanvitalia*   for the SanVitali family of Parma

*sapidus -a -um*   pleasant-tasted, flavoursome, savoury

*sapientium*   of the wise, of man (implies superiority, compare
   with *Troglodytarum*)

*Sapindus*   Indian-soap (from its use)

*saponaceus -a -um, saponarius -a -um*   lather-forming, soapy

*Saponaria*   Soap-like (lather-forming)

*sapota*   an old Greek name for a wild pear used by Linnaeus for
   the sapodilla or chicle tree

*saracenicus -a -um, sarracenicus -a -um*   of the Saracens

*Saracha*   for Isidore Saracha, a Benedictine monk

*sarachoides*   resembling *Saracha*

*sarc-, sarco-*   fleshy-

*Sarcocephalus*   Fleshy-head (the head of fruits)

*Sarcococca*   Fleshy-berry

*sarcodes*   flesh-like

*sardensis -is -e*   from Sardis, Smyrna, Asia Minor

*sardosus -a -um, sardous -a -um*   from Sardinia, Sardinian

*sarmaticus -a -um*   from Sarmatia on the Russo-Polish border

*sarmentaceus -a -um, sarmentosus -a -um*   with long slender
   stolons or runners

*sarniensis -is -e*   from Guernsey (Sarnia), Channel Isles

*saro-*   broom-like-

*Sarothamnus*   Broom-shrub

*Sarracenia*   for Dr D. Sarrasin who introduced *S. purpurea* from
   Quebec

*Sasa*   the Japanese name for certain bamboos

*sassafras*   from the Spanish name for saxifrage, 'salsafras'

*sativus -a -um*   planted, cultivated, sown, not wild

*Satureia, Satureja*   the Latin name for a culinary herb

*sauro-*   lizard-like-, lizard-

*Saussurea*   for the Swiss philosopher H. B. de Saussure

*savannarum*   of savannas

*saxatilis -is -e*   living in rocky places

*saxicolus -a -um, saxosus -a -um*   rock-dwelling

*Saxifraga*   Stone-breaker (lives in rock cracks and had a
   medicinal use for gall-stones)

*saxosus -a -um*   or rocky or stony places

*scaber -ra -rum, scabri*   scurfy, rough

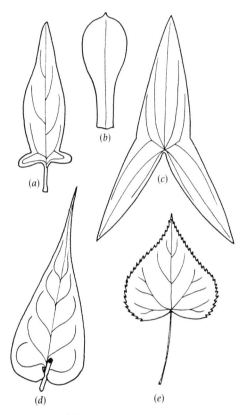

Fig. 6. More leaf shapes which provide specific epithets:
(a) hastate (e.g. *Scutellaria hastifolia* L.) with auricled leaf-base; (b) spathulate (e.g. *Sedum spathulifolium* Hook.); (c) sagittate (e.g. *Sagittaria sagittifolia* L.) with pointed and divergent auricles; (d) amplexicaul (e.g. *Polygonum amplexicaule* D. Don) with the basal lobes of the leaf clasping the stem; (e) cordate (e.g. *Tilia cordata* Mill.), heart-shaped.

*scaberulus -a -um, scabriusculus -a -um*   roughish, somewhat rough
*Scabiosa*   Scabies (former medicinal treatment for)
*scabrosus -a -um*   rather rough

170

*scalaris -is -e*   ladder-like

*scandens*   climbing

*scandicus -a -um*   from Scandia, Scandinavian

*Scandix*   ancient name for Shepherd's needle

*scaphi-, scapho-, scaphy-*   boat-shaped-, bowl-shaped-

*scapi-*   clear-stemmed-

*scapiger -era -erum*   scape-bearing

*scaposus -a -um*   with scapes or leafless flowering stems

*scardicus -a -um*   from Sar Planina (Scardia), south Yugoslavia

*scariola* (*serriola*)   endive-like, of salads

*scariosus -a -um*   shrivelled, thin, not green, scarious

*scarlatinus -a -um*   bright-red

*sceleratus -a -um*   of vile places, vicious, wicked (causes ulceration)

*sceptrum*   of a sceptre

*Scheuchzeria*   for the brothers J. & J. J. Scheuchzer, Swiss botanists

*schinseng*   from the Chinese name

*Schinus*   from the Greek name for another mastic-producing plant (*Pistacia*)

*schis-, schismo-, schiz-, schizo-*   divided-, cut-

*schistosus -a -um*   slate-coloured

*Schizanthus*   Divided-flower (the lobes of the corolla in the poor man's orchid)

*Schizaea*   Cut (the incised fan-shaped fronds)

*Schkuhria*   for Christian Schkuhr, German botanist (1741–1811)

*schoeno-*   rush-like, resembling *Schoenus*

*Schoenoplectus*   Rush-plait

*schoenoprasus -a -um*   rush-like leek (the leaves)

*Schoenus*   the old name for rush-like plants

*sciadi-, sciado-*   shade-, shaded-, canopy-, umbelled-

*sciaphilus -a -um*   shade-loving

*Scilla*   the ancient Greek name for the squill *Urginea maritima*

*scilloides*   squill-like, resembling *Scilla*

*scintillans*   sparkling, gleaming

*sciophilus -a -um*   shade-loving

*Scirpus*   the old name for a rush-like plant

*scitulus -a -um*   neat, pretty

*scitus -a -um*   fine

*sciuroides*   squirrel-tail-like

*sclareus -a -um*   clear (from a name for a *Salvia* used for eye lotions, clary)

*Scleranthus*   Hard-flower (texture of the perianth)

*sclero-*   hard-

*Scleropoa*   Hard-pasturage

*Scolopendrium*   Dioscorides' name for the hart's tongue fern compares the numerous sori to the legs of a millipede

*Scolymus*   the ancient Greek name for *S. hispanicus* and its edible root

*scoparius -a -um*   broom-like (use for making twig-brushes)

*scopulinus -a -um*   twiggy

*scopulorum*   of cliffs and rock faces

*scorbiculatus -a -um*   with surface depressions or grooves

*Scordium*   Dioscorides' name for a plant with the smell of garlic

*scorodonia*   an old generic name for garlic

*scorodoprasum*   a name used by Dioscorides for a garlic-like leek (has intermediate features)

*scorpioides*   curved like a scorpion's tail (see Fig. 3), scorpion-like

*scorteus -a -um*   leathery

*Scorzonera*   derivation uncertain but generally thought to refer to use as an antifebrile in snakebite

*scoticus -a -um*   from Scotland, Scottish

*scotinus -a -um*   dusky, dark

*scottianus -a -um*   for Munro B. Scott or Robert Scott of Dublin (1757–1808)

*scriptus -a -um*   marked with lines which suggest writing

*Scrophularia*   Scrophula (the glands on the corolla)

*scrotiformis -is -e*   shaped like a small double bag, pouch-shaped

*sculptus -a -um*   carved

*scutatus -a -um*   like a small round shield or buckler

*Scutellaria*   Dish (the depression of the fruiting calyx)

*scutellatus -a -um*   shield-shaped, platter-like

*scypho-*   cup-, beaker-

*se-*   apart-, without-, out-

*sebaceus -a -um, sebifer -era -erum*   producing wax

*sebosus -a -um*   full of wax

*Secale*   the Latin name for a grain-like rye

*secalinus -a -um*   rye-like, resembling *Secale*

*sechellarus -a -um*   from the Seychelles, Indian Ocean

*seclusus -a -um*   hidden, secluded

*secundi-, secundus -a -um*   turned-, one-sided (as when flowers are all to one side of an inflorescence) secund

*securiger -era -erum*   axe-bearing (the shape of some organ)

*Securinega*   Axe-refuser (the hardness of the timber)

*sedi-, sedioides*   stonecrop-like, resembling *Sedum*

*Sedum*   a name in Pliny (refers to the plant's 'sitting' on rocks etc. in the case of cushion species)

*segetalis -is -e, segetus -a -um*   of cornfields

*Selaginella*   a diminutive of *selago* (see below)

*selaginoides*   clubmoss-like, resembling *Selaginella*

*selago*   the name in Pliny for *Lycopodium*, from the Celtic name for the Druidic plant *Juniperus sabina*

*seleni-*   moon-

*Selinum*   the name in Homer for a celery-like plant with lustrous petals (relates etymologically with *Silaum* and *Silaus*)

*selliformis -is -e*   saddle-shaped, with both sides hanging down (e.g. of leaves)

*selloi, sellovianus -a -um, sellowii, seloanus -a -um*   for Friedrich Sellow (Sello), German botanist (1789–1831)

*semi-*   half-

*semidecandrus -a -um*   with (about) five stamens

*semipersistens*   half-persistent

*semiteres*   half-cylindrical

*semper-*   always-, ever-

*semperflorens*   ever-flowering

*sempervirens*   always green

*sempervivoides*   houseleek-like, resembling *Sempervivum*

*Sempervivum*   never-die, always living, always alive

*senanensis -is -e*   from Senan, China

*Senecio*   Old-man (the name in Pliny refers to the grey hairiness as soon as fruiting commences)

*senescens*   turning hoary with whitish hairs

*senilis -is -e*   aged, grey-haired

*sensibilis -is -e, sensitivus -a -um*   sensitive to touch

*senticosus -a -um*   thorny, full of thorns

*sepiarius -a -um, sepium*   growing in hedges, of hedges

*sepincolus -a -um*   hedge-dweller, inhabiting hedges

*sept-*   seven-

*septemfidus -a -um*   with seven divisions, seven-cut

*septentrionalis -is -e*   of the north, northern

*sepulchralis -is -e*   of tombs, of graveyards

*Sequoia*   for the Indian Sequoiah who invented the Cherokee alphabet (1770–1843)

*Sequoiadendron*   *Sequoia*-tree (resemblance in size)

*serapias*   an ancient name for an orchid

*seri-, serici-, sericans, sericeus -a -um*   silky-hairy (sometimes implying Chinese)

*serialis -is -e*   -rowed

*sericifer -era -erum, sericofer -era -erum*   silk-bearing

*-seris*   -potherb

*serissimus -a -um*   very late

*serotinus -a -um*   of late (flowering or fruiting) season

*serpens, serpentinus -a -um*   creeping, serpentine

*serpyllifolius -a -um*   thyme-leaves

*serra-*   saw-, saw-like-, serrate-

*Serrafalcus*   for the Duke of Serrafalco, archaeologist

*Serratula*   Saw-tooth (the name in Pliny for betony)

*serratus -a -um*   edged with forward pointing teeth (see Fig. 4(c)), serrate

*serriolus -a -um*   in ranks, of salad (from an old name for chicory)

*serrulatus -a -um*   edged with small teeth, finely serrate, serrulate

*Sesamum*   from the Semitic name

*Sesbania*   from an Arabic name

*Seseli*   the ancient Greek name

*Sesleria*   for Leonardo Sesler of Venice (d. 1785)

*sesqui-*   one-and-one-half-

*sesquipedalis -is -e*   about 18 inches long, the length of a foot and a half

*sesseli-, sessilis -is -e*   attached without a distinct stalk, sessile

*sessilifolius -a -um*   leaves without petioles, sessile-leaved

*setaceus -a -um, seti-*   with bristles or stiff hairs, bristly

*Setaria*   Bristly (the hairs subtending the spikelets)

*setifer -er -erum, setiger -era -erum*   bearing bristles, bristly

*setispinus -a -um*   bristle-spined

*setosus -a -um*   covered with bristles or stiff hairs

*setulosus -a -um*   slightly bristly

*sex-*   six-

*sexangularis -is -e*   six-angled (stems)

*shallon*   from a Chinook Indian name

*Sherardia*   for William (1659–1728) or James Sherard

*siameus -a -um*   from Thailand (Siam)

*Sibbaldia*   for Prof. Robert Sibbald of Edinburgh (1643–1720)

*sibiricus -a -um*   from Siberia, Siberian

*Sibthorpia*   for John Sibthorp, English botanist (1758–1790) and his son Humphrey

*siccus -a -um*   dry

*siculi-*   dagger-shaped-

*siculus -a -um*   from Sicily, Sicilian

*Sida*   from a Greek name for a water-lily
*sidereus -a -um*   iron-hard
*Sideritis*   the Greek name for plants used to dress wounds
   caused by iron weapons
*Sideroxylon*   Iron-wood (the hard timber of the miraculous
   berry)
*sieboldiana, sieboldii*   for Philipp Franz von Siebold, collector in
   Japan (1796–1866)
*Sieglingia*   for Prof. Siegling of Erfurt
*Sigesbeckia* (*Siegesbeckia, Sigesbekia*)   for J. G. Sigesbeck of
   Leningrad
*signatus -a -um*   well-marked, designated, signed
*sikkimensis -is -e*   from Sikkim, Himalayas
*Silaum*   meaning uncertain (see *Selinum*)
*Silaus*   an old generic name in Pliny used for pepper saxifrage
*Silene*   Theophrastus' name for *Viscaria*, another catchfly
*siliceus -a -um*   growing on sand
*siliculosus -a -um, siliquosus -a -um*   having elongate pods
*siliquastrum*   the old Latin name for a pod-bearing tree
*silvaticus -a -um, silvester -tris -tre*   of woodlands, of woods, wild
*Silybum*   Dioscorides' name for a thistle-like plant
*Simarouba* (*Simaruba*)   from the Carib name for bitter damson
*simensis -is -e*   from Arabia, Simenia, middle eastern
*Simethis*   after Acis' mistress, the nymph Simethis
*simia*   of the ape or monkey (flower-shape or implying
   inferiority)
*similis -is -e*   resembling other species, like, similar
*simplex, simplici-*   undivided, simple
*simulans, simulatus -a -um*   resembling, imitating, similar
*Sinapis*   the old name used by Theophrastus for mustard
*sinensis -is -e* (*chinensis -is -e*)   from China, Chinese
*singularis -is -e*   unusual, singular
*sinicus -a -um, sino-*   of China, Chinese
*sinistrorsus -a -um*   twining clockwise upwards (as seen from
   outside)
*sino-*   Chinese-
*Sinowilsonia*   for E. H. Wilson, introducer of Chinese plants
   (1876–1930)
*sinuatus -a -um, sinuosus -a -um*   waved, with a wavy margin
   (see Fig. 4(*e*)), sinuate
*siphiliticus -a -um*   of syphilis, used to treat the disease *Lobelia
   siphilitica*
*sipho-, -siphon*   tubular-, -pipe, -tube

*sipyleus*   flour or meal
*sisalanus -a -um*   from Sisal, Yucatan, Mexico
*sisara*   Dioscorides' name for a plant with an edible root
*Sison*   a name used by Dioscorides
*Sisymbrium*   ancient Greek name for various plants
*Sisyrinchium*   Theophrastus' name for an iris
*sitchensis -is -e*   from Sitka Island, Alaska
*Sium*   an old Greek name for water plants
*Skimmia*   from a Japanese name
*smaragdinus*   of emerald
*Smilacina*   diminutive of *Smilax*
*Smilax*   from an ancient Greek name
*Smyrnium*   myrrh-like (the fragrance)
*sobolifer -era -erum*   producing vigorous shoots from the stem at
    ground level
*socialis -is -e*   growing in colonies
*sodomeus -a -um*   from the Dead Sea area (Sodom)
*solan-, solani-*   potato-, *Solanum*-like-
*Solanum*   Comforter (a name in Pliny)
*solaris -is -e*   of the sun, of sunny habitats
*Soldanella*   Coin-shaped (the leaves)
*solen-, soleno-*   box-, tube-
*Solenostemon*   Tube-stamens (their united filaments)
*Solidago*   Uniter (used as a healing medicine)
*solidifolius -a -um*   entire-leaved
*solidus -a -um*   a coin, solid, dense, not hollow
*solstitialis -is -e*   of mid-summer (flowering-time)
*somnians*   asleep, sleeping
*somnifer -era -erum*   sleep-inducing, sleep-bearing
*Sonchus*   the Greek name for a thistle
*sophia*   wisdom (the use of flixweek in healing)
*soporificus -a -um*   sleep-bringing, soporific
*Sorbus*   the Latin name for the service tree
*sordidus -a -um*   neglected, dirty-looking
*Sorghum*   from the Italian name
*sororius -a -um*   very closely related, sisterly
*soulangiana, soulangii*   for Etienne Soulange-Bodin, French
    horticulturalist (1744–1846)
*spadiceus -a -um*   chestnut-brown, date-coloured
*Sparaxis*   Tear (the torn bracts)
*Sparganium*   Dioscorides' name for bur-reed
*sparsi-, sparsus -a -um*   scattered
*Spartina*   old name for various plants used to make ropes

*Spartium*   Binding or Broom (former use for binding and
   sweeping)
*spathi-, spatho-*   spathe- (as in arums)
*Spathodea*   Spathe-like (the calyx)
*spathulatus -a -um, spathuli-*   shaped like a spoon, spathulate
   (see Fig. 6(*b*))
*spatiosus -a -um*   spacious, wide
*speciosus -a -um*   showy, handsome
*spectabilis -is -e*   admirable, spectacular
*specularius -a -um, speculum*   shining, mirror-like
*speculatus -a -um*   shining, as if with mirrors
*speluncae, speluncarum*   of caves, cave-dwelling
*speluncatus -a, -um, speluncosus -a -um*   cavities, full of holes
*Spergula*   Scatterer (l'Obel's name refers to the discharge of the
   seeds)
*Spergularia*   resembling *Spergula*
*-spermus -a -um*   -seeded
*sphacelatus -a -um*   necrotic, gangrened
*sphaer-, sphaero-*   globular-, spherical-
*sphaerocephalus -a -um*   round-headed
*sphegodes*   resembling wasps (flower shape)
*spheno-*   wedge-
*sphondilius -a -um*   rounded
*spica, spicati-, spicatus -a -um, spicifer -era -erum*   with an
   elongate inflorescence of sessile flowers, spiked, spicate (see
   Fig. 2(*a*))
*spicant*   tufted (spikenard, spike, ear)
*spica-venti*   ear of the wind, tuft of the wind
*spiculi-*   spicule-, small-thorn-
*Spigelia*   for Adrian van der Spiegel of Padua (1578–1625)
*Spilanthes*   Stained-flower (receptacular marks of some species)
*spina-christi*   Christ's thorn
*spinescens, spinifer -era -erum, spinifex, spinosus -a -um*   spiny,
   with spines
*spinulifer -era -erum, spinulosus -a -um*   with small spines
*Spiraea*   Theophrastus' name for a plant used for making
   garlands
*spiralis -is -e*   twisted, spiral
*Spiranthes*   Twisted (the inflorescence)
*splendens, splendidus -a -um*   gleaming, striking
*Spondias*   Theophrastus' name refers to the plum-like fruit
*spongiosus -a -um*   spongy
*sponhemicus -a -um*   from Sponheim, Rhine

*Sporobolus*   Seed-caster

*-sporus -a -um*   -seed, -seeded

*spretus -a -um*   spurned

*spumarius -a -um*   foamy, frothing

*spurius -a -um*   false, bastard, spurious

*squalens, squalidus -a -um*   untidy, dingy, squalid

*squamarius -a -um*   scale-clad, covered with scales

*squamatus -a -um*   with small scale-like leaves or bracts (squamae), squamate

*squamigerus -era -erum*   scale-bearing

*squamosus -a -um*   very scaly, full of scales

*squarrosus -a -um*   rough (as when small overlapping leaves have protruding tips) spreading in all directions

*stachy-*   spike-like-, resembling *Stachys*

*Stachys*   Spike (the Greek name for several dead-nettles)

*-stachyon, -stachys, stachyus -a -um*   -spikeleted, -panicled

*Stachytarpheta*   Thick-spike

*stagnalis -is -e*   of pools

*stagninus -a -um*   of swampy or boggy ground

*stamineus -a -um*   with prominent stamens

*stans*   upright, erect, standing

*Stapelia*   by Linnaeus for J. B. von Stapel, Dutch physician of Amsterdam

*Staphylea*   Cluster (a name in Pliny, refers to the flowers)

*-staphylos*   -bunch (as of grapes)

*Statice*   Astringent (Dioscorides' name for the *Limonium* of gardeners)

*stauro-*   cross-shaped-, crosswise-, cruciform-

*steiro-*   barren-

*Stellaria*   Star (the appearance of stitchwort flowers)

*stellaris -is -e, stellatus -a -um*   star-like, with spreading rays, stellate

*stelliger -era -erum*   star-bearing

*stellipilus -a -um*   with stellate hairs

*-stemon*   -stamened

*sten-, steno-*   narrow-

*stenopetalus -a -um*   narrow-petalled

*Stenotaphrum*   Narrow-trench (the florets are recessed into cavities in the rachis)

*stephan-, stephano-*   crown-

*Stephanotis*   Crown (the auricled staminal crown) also used by the Greeks for other plants used for making chaplets or crowns

*stepposus -a -um*   of the Steppes

*sterilis -is -e*   infertile, with barren fruit, sterile

*Sternbergia*   for Count Kaspar Moritz von Sternberg of Prague (1761–1838)

*-stichus -a -um*   -ranked, -rowed

*stict-, sticto-, -stictus -a -um*   punctured-, -spotted

*stictocarpus -a -um*   with spotted fruits

*stigmaticus -a -um*   spotted, dotted, marked

*stigmosus -a -um*   spotted

*Stilbe*   Shining

*Stipa*   Tow (Greek use of the feathery inflorescences, like hemp, for caulking and plugging

*stipellatus -a -um*   with stipels (in addition to stipules)

*stipitatus -a -um*   with a stipe or stalk

*stipulaceus -a -um, stipularis -is -e, stipulatus -a -um, stipulosus -a -um*   with conspicuous stipules

*stoechas*   Dioscorides' name for a lavender grown on the Iles d'Hyères, Toulon, which were called 'Stoichades'

*stolonifer -era -erum*   spreading by stolons, with creeping stems which root at the nodes

*stragulus -a -um*   carpeting, mat-forming

*stramine- stramineus -a -um*   straw-coloured

*stramonium*   spiky-fruit (used by Theophrastus as a name for the thorn apple)

*strangulatus -a -um*   constricted, strangled

*Stranvaesia*   for W. T. H. F. Strangways, Earl of Ilchester (1795–1865)

*Stratiotes*   Soldier (Dioscorides' name for an Egyptian water plant)

*Strelitzia*   for Charlotte of Mecklenburg-Strelitz, wife of George III

*strepens*   rustling, rattling

*strept-, strepto-*   twisted-, coiled-

*Streptocarpus*   Twisted-fruit

*striatellus -a -um, striatulus -a -um, striatus -a -um*   marked with parallel lines, grooves or ridges, striated

*stricti-, stricto-, strictus -a -um*   straight, erect, strict

*Striga, strigatus -a -um*   Swathe, straight, rigid, *Striga*-like

*strigilosus -a -um*   with short appressed bristles

*strigosus -a -um*   with short appressed bristles

*strigosus -a -um*   with rigid hairs or bristles, strigose

*strigulosus -a -um*   somewhat strigose

*striolatus -a -um*   faintly striped

*strobilaceus -a -um*   cone-like, cone-shaped
*strobilifer -era -erum*   cone-bearing
*strobus*   an ancient name for an incense-bearing tree
*strongyl-, strongylo-*   round-, rounded-
*Strophanthus*   Twisted-flower (the elongate lobes of the corolla)
*strumarius -a -um, strumosus -a -um*   cushion-like, swollen
   (refers to former medicinal use for the treating of swollen
   necks)
*Struthiopteris*   Ostrich-feather (the fertile fronds)
*Strychnos*   Linnaeus reapplied Theophrastus' name for
   poisonous solanaceous plants
*stygia*   of the underworld, Stygean (*Globularia stygia* spreads by
   subterranean stolons)
*stylosus -a -um*   with a prominent style
*-stylus -a -um*   -styled
*styracifluus -a -um*   flowing with gum
*Styrax*   ancient Greek name for storax gum tree
*Suaeda*   from the Arabic name
*suaveolens*   sweet-scented
*suavis -is -e*   sweet, agreeable
*sub-, suc-, suf-, sug-*   below-, under-, approaching-, nearly-,
   just-, less than-, usually-
*subacaulis -is -e*   almost without a stem
*subcaeruleus -a -um*   slightly blue
*suber*   corky (bark of the cork oak)
*suberosus -a -um*   slighly bitten, corky
*sublustris -is -e*   glimmering, almost shining
*submersus -a -um*   under-water, submerged
*subsessilis -is -e*   very short stalked, almost-sessile
*subterraneus -a -um*   below ground, undergound
*subtilis -is -e*   fine
*Subularia*   Awl (the leaf shape)
*subulatus -a -um*   awl-shaped, subulate
*Succisa*   Cut-off (the rhizome of *S. pratensis*)
*succisus -a -um*   cut off from below
*succosus -a -um*   full of sap, sappy
*succotrinus -a -um*   from Socotra, Indian Ocean
*Succowia*   for Georg Adolph Suckow of Heidelberg (1751–1813)
*succulentus -a -um*   fleshy, soft, juicy, succulent
*sudanensis -is -e*   from the Sudan, Sudanese
*sudeticus -a -um*   from the Sudetenland of Czechoslovakia and
   Poland
*suecicus -a -um*   from Sweden, Swedish

*Sueda*   from the Arabic for salt

*suffocatus -a -um*   suffocating (the flower heads turn to the ground)

*suffruticosus -a -um*   somewhat shrubby at the base, soft-wooded and growing yearly from ground level

*suionum*   of the Swedes (Sviones)

*sulcatus -a -um*   furrowed, grooved, sulcate

*sulfureus -a -um, sulphureus -a -um*   pale-yellow, sulphur-yellow

*sultani*   for the Sultan of Zanzibar

*sumatranus -a -um*   from Sumatra, Indonesia

*super-, supra-*   above-, over-

*superbus -a -um*   magnificent, proud, superb

*superciliaris -is -e*   eyebrow-like, with eyebrows, with hairs above

*supinus -a -um*   lying flat, extended, supine

*supranubius -a -um*   of very high mountains, from above the clouds

*surattensis -is -e*   from Surat on the West coast of India

*surculosus -a -um*   shooting, suckering

*susianus -a -um*   from Susa, Iran

*suspendus -a -um, suspensus -a -um*   hanging down, pendent, suspended

*sutchuensis -is -e*   from Szechwan, China

*sy-, syl-, sym-, syn-, syr-, sys-*   with-, together with-, united-, joined-

*sycamorus*   fig-fruited, of the fig

*sylvaticus -a -um, sylvester -tris -tre*   wild, of woods or forests, sylvan

*sylvicola*   inhabiting woods

*sympho-, symphy-*   growing-together-

*Symphoricarpus(os)*   Clustered-berries

*Symphytum*   Live-together (Dioscorides' name for healing plants, including comfrey, *conferva* of Pliny)

*syphiliticus -a -um*   (see siphiliticus)

*syriacus -a -um*   from Syria, Syrian

*Syringa*   Pipe (use of the hollow stems to make flutes)

*syzigachne*   with scissor-like glumes

*Syzygium*   Paired (from the form of branching and opposite leaves. Formerly applied to *Calyptranthus*)

*Tabebuia*   from a Brazilian name

*tabernaemontanus -a -um*   for J. T. Bergzabern of Heidelberg

*tabulaeformis -is -e, tabuliformis -is -e*   flat and circular, plate-like *tabularis -is -e, tabuli-*  table-flat, flattened

*tacamahacca*  from an Aztec name

*tacazzeanus -a -um*  from the Takazze River, Ethiopia

*Tacca*  from a Malayan name for arrowroot

*Taccarum*  Tacca-arum (implies intermediate looks but not hybridity)

*taccifolius -a -um*  with leaves like *Tacca*

*taeda*  an ancient name for resinous pine trees

*taediger -era -erum*  torch-bearing

*taenianus -a -um*  shaped (segmented) like a tapeworm

*taeniosus -a -um*  ribbon-like, banded(-leaves)

*Tagetes*  for Tages, Etruscan god and grandson of Jupiter

*taiwanensis -is -e*  from Formosa (Taiwan), Formosan

*tamarici-, tamarisci-*  tamarisk-like-

*Tamarindus*  Indian-date (from the Arabic 'tamr')

*Tamarix*  the Latin name, perhaps from the Spanish area of the River Tambo (Tamaris)

*tamnifolius -a -um*  bryony-leaved, with leaves like *Tamus* (Tamnus of Pliny)

*Tamus*  from the name in Pliny for a kind of vine

*Tanacetum*  Immortality (tansy was placed amongst the winding sheets of the dead to repel vermin)

*tanaciti-*  tansy-like-

*tanguticus -a -um*  of the Tangut tribe of north-east Tibet, Tibetan

*tapein-, tapeino-*  humble-, modest-

*tapeti-*  carpel-like-

*taraxaci-*  dandelion-like-

*Taraxacum*  Disturber (from the Persian name for a bitter herb)

*tardi-, tardivus -a -um, tardus -a -um*  slow, reluctant, late

*tartareus -a -um*  of the underworld (coloration), with a loose crumbling surface

*tartaricus -a -um, tataricus -a -um*  from Tartary, central Asia

*tauricus -a -um*  from the Crimea (Tauria)

*taurinus -a -um*  from Turin, Italy, or of bulls

*taxi-*  yew-like-, resembling *Taxus*-

*taxodioides*  resembling *Taxodium*

*Taxodium*  Yew-like, resembling-*Taxus*

*taxoides*  resembling yew

*Taxus*  the Latin name of yew

*tazetta*  little cup (the corona of *Narcissus tazetta*)

*technicus -a -um*  special, technical

*Tecoma, Tecomaria*  from the Mexican name of the former

*Tectona*   from the Tamil name for teak 'tekka'

*tectorum*   of tiles, of rooftops, growing on rooftops

*tectus -a -um*   with a thin covering, tectate

*Teesdalia*   for Robert Teesdale, Yorkshire botanist

*tef*   the Arabic name for *Eragrostis abyssinica* (tef grass)

*tegetus -a -um*   mat-like

*tel-, tele-*   far-, far-off-

*telephioides*   resembling *Sedum telephium*

*Telephium*   a Greek name for a plant thought to be capable of indicating reciprocated love (far-off-lover)

*Tellima*   an anagram of *Mitella*

*telmataia, telmateius -a -um*   of marshes, of muddy water

*telonensis -is -e*   from Toulon (Telenium), France

*temenius -a -um*   of sacred precincts or holy places

*temulentus -a -um, temulus -a -um*   bewildered, intoxicated, drunken (toxic seed of ryegrass)

*tenax*   gripping, stubborn, firm, persistent, tenacious

*tenebrosus -a -um*   delicate, somewhat tender

*tenellus -a -um, teneri-, tener -era -erum*   soft, tender, delicate

*tenens*   enduring, persisting

*tenui-, tenuis -is -e*   slender, thin, fine

*tenuifolius -a -um*   slender-leaved

*tenuior*   more slender

*tephro-*   ash-grey-

*Tephrosia*   Ashen (the leaf colour)

*terebinthi-*   pistacia-like-, turpentine-

*terebinthifolius -a -um*   with leaves like those of *Pistacia terebinthus*

*terebinthinus*   a former name for Chian turpentine tree, *Pistacia terebinthus*

*teres teretis terete, tereti-*   quill-like, cylindrical, terete

*tereticornis -is -e*   with cylindrical horns

*teretiusculus -a -um*   somewhat smoothly rounded

*terminalis -is -e*   terminal (the flower on the stem)

*ternateus -a -um*   from the Ternate Islands, Moluccas

*ternatus -a -um, ternati-, terni-*   with parts in threes, ternate (see Fig. 5(*e*))

*terrestris -is -e*   growing on the ground, not epiphytic or aquatic

*tessellatus -a -um*   chequered, mosaic-like, tessellated

*testaceus -a -um*   brownish-yellow

*testicularis -is -e, testiculatus -a -um*   tubercled, having some testicle-shaped structure (e.g. a tuber)

*testudinarium*   resembling tortoise shells

*tetra-*   four-

*Tetracme*   Four-points (the shape of the fruit)

*Tetragonolobus*   Quadrangular-pod (the fruit)

*tetragonus -a -um*   four-angled

*tetrahit*   four-time (tetraploid), foetid

*tetralix*   a name used by Theophrastus for the cross-leaved state when the leaves are arranged in whorls of four

*tetrandrus -a -um*   with four stamens, four-anthered

*tetraplus -a -um*   fourfold (e.g. ranks of leaves)

*Tetraspis*   Four shields

*Teucrium*   Dioscorides' name perhaps for Teucer, hero and first King of Troy

*texanus -a -um, texensis -is -e*   from Texas, USA, Texan

*textilis -is -e*   used for weaving

*thalassicus -a -um*   growing in the sea

*Thalia, thalianus -a -um*   for Johann Thal, German botanist (1542–1583)

*Thalictrum*   a name used by Dioscordes

*-thamnus -a -um*   -shrub-like, -shrubby

*Thapsia*   ancient Greek name used by Theophrastus

*Thapsus*   from the Island of Thapsos

*thebaicus -a -um*   from Thebes, Greece

*theco-, -thecus -a -um*   box-, -chambered

*theifer -era -erum*   tea-bearing

*thele-, thelo-, thely-*   female-, nipple-

*Thelycrania*   the name used by Theophrastus

*Thelygonum*   Girl-begetter (claimed by Pliny to cause girl offspring)

*Thelypteris*   Female-fern

*Themeda*   from an Arabic name

*Theobroma*   God's-food

*theriacus -a -um*   of snakes (medicinal use in snakebite)

*thermalis -is -e*   of warm springs

*Thermopsis*   Lupin-like

*thero-*   summer-

*Thesium*   a name in Pliny for a bulbous plant

*Thespesia*   Divine (commonly cultured round temples)

*thessalonicus -a -um, thessalus -a -um*   from Thessaly

*Thevetia*   for André Thévet, French traveller in Brazil and Guiana (1502–1592)

*thibeticus -a -um*   from Tibet

*thirsi-*   panicled-

*Thladiantha*   Eunoch-flower (female flowers have aborted stamens)

*Thlaspi*   the name used by Dioscorides for cress of corruption, of ruination (a medieval name for a poisonous buttercup)

*Thrinax, -thrinax*   fan, -fanned, -trident

*-thrix*   -hair, -haired

*Thuja, Thuya*   Theophrastus' name for a resinous fragrant-wooded tree

*Thujopsis, Thuyopsis*   resembling *Thuja*

*Thunbergia*   for K. P. Thunberg of Uppsala (1743–1822)

*thurifer -era -erum*   incense-bearing

*thuringiacus -a -um*   from mid-Germany (Thuringia)

*thuyioides, thyoides*   *Thuja*-like

*thymbra*   an ancient Greek name for a savory thyme-like plant

*Thymelaea*   Thyme-olive (the leaves and fruit)

*Thymus*   Theophrastus' name for a plant used in sacrifices

*thyrsi-, thyrsoideus -a -um*   panicle-like, thyrsoid (see Fig. 3(*d*))

*thysano-, thysanoto-*   fringed-

*Tiarella*   Little-diadem (the capsules)

*tibeticus -a -um*   from Tibet

*tibicinis*   piper's or flute-player's

*tibicinus -a -um*   flute-like

*Tibouchina*   from a Guianan name

*Tigridia*   Tiger (the markings of the perianth)

*tigrinus -a -um*   striped, spotted, tiger-toothed

*Tilia*   Wing (the Latin name for the lime tree)

*tiliae-, tiliaceus -a -um*   lime-like, resembling *Tilia*

*Tillandsia*   for Elia Tillands, Swedish botanist

*tinctorius -a -um*   used for dyeing

*tinctorum*   of the dyers

*tinctus -a -um*   coloured

*tingens*   stained, dyed

*tingitanus -a -um*   from Tangiers, Morocco

*tinus*   the old Latin name for laurustinus (*Viburnum*)

*tipuliformis -is -e*   resembling a Tipulid (crane fly)

*tirolensis -is -e*   from the Tyrol, Tyrolean

*titano-*   chalk-, lime-

*titanum, titanus -a -um*   of the Titans, gigantic, very large

*Tithonia*   after a character from Greek mythology

*tithymaloides*   spurge-like

*Tithymalus*   an ancient name for plants with latex, spurges

*Todea*   for H. J. Tode, German mycologist (1733–1797)

*Tofieldia*    for Thos. Tofield, Yorkshire naturalist (1730–1779)
*togatus -a -um*    robed, gowned
*Tolmiea*    for Dr W. F. Tolmie of the Hudson's Bay Company
*Tolpis*    a name of uncertain derivation
*tomentellus -a -um*    somewhat hairy
*tomentosus -a -um*    thickly matted with hairs
*tonsus -a -um*    shaven, sheared, shorn
*Tordylium*    the name used by Dioscorides
*Torilis*    a meaningless name by Adanson
*toringoides*    toringo-like (Japanese name for a *Malus*)
*torminalis -is -e*    of colic (used medicinally to relieve colic)
*torminosus -a -um*    causing colic
*torosus -a -um*    cylindrical with regular constrictions
*Torreya*    for John Torrey, American botanist (1796–1873)
*torridus -a -um*    of very hot places
*tortilis -is -e, tortus -a -um*    twisted
*tortuosus -a -um*    meandering (irregularly twisted stems)
*torulosus -a -um*    swollen or thickened at intervals
*torvus -a -um*    fierce, harsh, sharp
*toxicarius -a -um, toxicus -a -um*    poisonous
*Toxicodendron(um)*    Poison-tree
*toza*    from a South African native name
*trachelium*    neck (old name for a plant used for throat
    infections)
*trachelo-*    neck-
*trachy-*    shaggy-, rough-
*trachyodon*    short-toothed, rough-toothed
*Trachystemon*    Rough-stamens
*Tradescantia*    for John Tradescant, gardener to Charles I (son of
    Old John Tradescant)
*tragacantha*    yielding gum-tragacanth (from a Greek plant name
    – goat-thorn)
*trago-*    goat-
*Tragopogon*    Goat-beard (the pappus of the fruit)
*Tragus*    Goat
*trans-*    through-, beyond-, across-
*transiens*    intermediate, passing-over
*translucens*    almost transparent
*transwallianus -a -um*    from Pembroke, south Wales (beyond
    Wales)
*transylvanicus -a -um*    from Romania (Transylvania)
*Trapa*    from *calcitrapa*, a four-spiked weapon used in battle to
    maim cavalry horses

*tremuloides*   aspen-like, resembling *Populus tremula*
*tremulus -a -um*   trembling, shaking
*tri-*   three-
*triacanthos, triacanthus -a -um*   three-spined
*triandrus -a -um*   three-stamened
*triangulari-, triangularis -is -e*   three-angled, triangular
*trich-, trichus -a -um*   hair-like-, -hairy
*Trichomanes*   Hair-madness (Theophrastus' name for maidenhair spleenwort)
*Trichophorum*   Hair-carrier (perianth bristles)
*Trichosanthes*   Hair-flower (the fringed corolla lobes)
*trichospermus -a -um*   hairy-seeded
*trichotomus -a -um*   three-forked, triple-branched
*tricoccus -a -um*   three-seeded, three-berried
*tricolor*   three-coloured
*tricornis -is -e, tricornutus -a -um*   with three horns
*tridactylites*   three-fingered
*Tridax*   Three-toothed (Theophrastus' name for a lettuce, ligulate florets are three-fid)
*trientalis -is -e*   a third of a foot in length, about four inches tall
*trifasciatus -a -um*   three-banded
*trifidus -a -um*   divided into three, three-cleft
*Trifolium*   Trefoil (the name in Pliny for trifoliate plants)
*trifurcatus -a -um*   three-forked
*Triglochin*   Three-barbed (the fruits)
*Trigonella*   Triangle (the flower of fenugreek seen from the front)
*trigonus -a -um*   three-angled, with three flat faces and angles between them
*Trillium*   In-threes (the parts are conspicuously in threes, lily-like)
*trimestris*   of three months, maturing in three months
*Trinia*   for K. B. Trinius, Russian botanist (1778–1844)
*trionus -a -um*   three-coloured
*Tripleurospermum*   Three-ribbed-seed
*tripli-, triplo-*   triple-, threefold-
*triqueter, triquetrus -a -um*   three-edged, three-angled (stems)
*Trisetum*   Three-awns
*tristis -is -e*   bitter, sad, dull-coloured
*triternatus -a -um*   three times in threes (division of the leaves)
*Triticum*   the classical name for wheat
*tritifolius -a -um*   with polished leaves
*Tritonia*   Weathercock
*tritus -a -um*   in common use

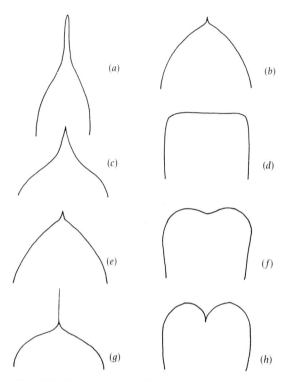

Fig. 7. Leaf-apex shapes which provide specific epithets:
(a) caudate (e.g. *Ornithogalum caudatum* Jacq.) with a tail; (b)
mucronate (e.g. *Erigeron mucronatus* DC.) with a hard tooth; (c)
acuminate (e.g. *Magnolia acuminata* L.) pointed abruptly; (d) truncate
(e.g. *Zygocactus truncatus* K. Schum.) bluntly foreshortened; (e)
apiculate (e.g. *Braunsia apiculata* Schw.) with a short broad point; (f)
retuse (e.g. *Daphne retusa* Hemsl.) shallowly indented; (g) aristate
(e.g. *Berberis aristata* DC.) with a hair-like tip, not always restricted
to describing the leaf-apex; (h) emarginate (e.g. *Limonium emargin-
atum* (Willd.) O. Kuntze) with a deep mid-line indentation.

*triumvirati*   of three men (like mayoral regalia)
*trivialis -is -e*   common, ordinary, wayside, of crossroads
*trixago*   *Trixis*-like
*Trixis*   Triple (three-angled fruits)

*trocho-*   wheel-like-, wheel-

*troglodytarum*   of apes or monkeys (implies inferiority or unsuitability for man, cf. *sapentium*

*Trollius*   from the Swiss-german 'Trollblume' – rounded flower

*Tropaeolum*   Trophy (the gardener's *Nasturtium* was likened by Linnaeus to the shields and helmets displayed after victories in battle)

*tropicus -a -um*   of the tropics

*truncatulus -a -um, truncatus -a -um*   blunt-ended (the apex of a leaf (see Fig. 7(*d*)), truncate

*Tsuga*   from the Japanese name for the hemlock cedar

*tubaeflorus -a -um*   with trumpet-shaped flowers

*tubatus -a -um*   trumpet-shaped

*Tuberaria*   Tuber (rootstock of *T. vulgaris*)

*tuberculatus -a -um, tuberculosus -a -um*   warted, warty, tuberculate (the surface texture)

*tubergenianus -a -um, tubergenii*   for the van Tubergen bulb growers of Holland

*tuberosus -a -um*   swollen, tuberous

*tubifer -era -erum, tubulosus -a -um*   tubular, bearing tubular structures

*tubiflorus -a -um*   with trumpet-shaped flowers

*tucumaniensis -is -e*   from Argentina, Argentinian

*tul-*   warted-

*Tulbaghia*   for Ryk Tulbagh, one time Governor of the Cape of Good Hope (1699–1771)

*Tulipa*   from the Persian name for a turban

*tulipi-*   tulip-, tulip-like-

*tulipiferus -a -um*   tulip-bearing, having tulip-like flowers

*tumescens*   inflated

*tumidi, tumidus -a -um*   swollen, tumid

*tunbrigensis -is -e*   from Tunbridge

*Tunica*   Undergarment (the bracts below the calyx)

*tunicatus -a -um*   coated, having a covering or tunic, tunicate

*tuolumnensis -is -e*   from Tuolumne river and county, USA

*tupi-, tupis-*   mallet-like-

*turbinatus -a -um*   top-shaped, turbinate

*turcicus -a -um*   from Turkey, Turkish

*turcomanicus -a -um*   from Turkestan

*turcumaniensis -is -e*   from Turku, Finland

*turfaceus -a -um*   growing in bogs

*turgidus -a -um*   inflated, turgid

*turgiphalliformis -is -e*   erect-phallus-shaped

*Turnera*   for William Turner, Tudor botanist of Wells, author of
  *A New Herbal* (1508–1568)

*turpis -is -e*   ugly, deformed

*Turrita, Turritis*   Tower

*Tussilago*   Coughwort (medicinal use of leaves for treatment of
  coughs)

*tylo-*   knob-, callus-, swelling-

*Typha*   a Greek name for various plants

*typhinus -a -um, typhoides*   bulrush-like, resembling *Typha*,
  relating to fever

*typicus -a -um*   the type, typical


*uber -is -e*   luxuriant, full, fruitful

*ucranicus -a -um*   from the Ukraine, Ukrainian

*udensis -is -e*   from the River Uda or the Uden district or
  Siberia

*uduensis -is -e*   from Udu, New Guinea

*ugandensis -is -e*   from Uganda, East Africa

*ugni*   from a Chilean name

*-ugo*   -having (a feminine suffix in generic names)

*ulcerosus -a -um*   knotty, lumpy

*-ulentus -a -um*   -abundant, -full

*Ulex*   a name in Pliny

*ulicinus -a -um*   resembling *Ulex*

*uliginosus -a -um*   marshy, of swamps or marshes

*-ullus -a -um*   -smaller, -lesser

*Ulmaria*   Elm-like (Gesner's name refers to the appearance of
  the leaves)

*ulmi-, ulmoides*   elm-like, resembling *Ulmus*

*Ulmus*   the Latin name for elms

*-ulus -a -um*   -tending to, -having somewhat

*ulvaceus -a -um*   resembling the green seashore alga *Ulva*

*umbellatus -a -um*   with the branches of the inflorescence all
  rising from the same point, umbellate (see Fig. 2(*e*))

*Umbellularia*   Little-umbel (the inflorescences)

*umbilicatus -a -um, umbilicus -a -um*   navelled, with a navel

*Umbilicus*   Navel (the depression in the leaf surface above the
  insertion of the petiole)

*umbo-*   knob-like-

*umbonatus -a -um*   with a raised central boss or knob

*umbracul-*   umbrella-like-

*umbraculiferus -a -um*   umbrella-bearing

*umbrosus -a -um*   growing in shade, shade-loving, giving shade

*Uncaria, uncatus -a -um, uncus -a -um*   Hooked, crooked, with hooked tips

*unctuosus -a -um*   with a smooth shiny surface, greasy

*undatus -a -um, undosus -a -um, undulatus -a -um*   not flat, wavy, waved, undulate

*unedo*   the Latin name for the *Arbutus* tree and its fruit meaning 'I eat one'

*ungui-*   clawed-

*unguicularis -is -e, unguiculatus -a -um*   with a small claw or stalk (e.g. the petals)

*uni-*   one-, single-

*uniflorus -a -um*   one-flowered

*unilateralis -is -e*   one-sided, unilateral

*unilocularis -is -e*   with a one-chambered ovary

*unioloides*   resembling *Uniola* (American sea oats)

*uplandicus -a -um*   from Uppland, Sweden

*uragogus -a -um*   diuretic

*uralensis -is -e*   from the Ural mountains, USSR

*urbanus -a -um, urbicus -a -um*   of towns

*urceolatus -a -um*   pitcher- or urn-shaped

*Urena*   from the Malabar name

*urens*   stinging, burning

*Urera*   Burning (cow itch)

*Urginea*   from the Algerian type locality, Beni Urgin

*urnigerus -a -um*   urn-bearing

*uro-, -urus -a -um*   tail-, -tailed

*Ursinia*   for J. Ursinus, author of *Arboretum Biblicum* (1608–1666)

*ursinus -a -um*   bear-like (the smell), northern (under the *Ursa major* constellation

*Urtica*   Sting (the Latin name)

*Urvillea*   see *Durvillaea*

*-usculus -a -um*   -ish (a diminutive ending)

*usitatissimus -a -um*   everyday, ordinary, most useful

*Usnea*   a name of uncertain meaning by Adanson

*usneoides*   resembling *Usnea*, hanging in long threads

*ustulatus -a -um*   scorched-looking

*utilis -is -e*   useful

*Utricularia*   Little-bottle (the underwater traps of the bladderwort)

*utricularis -is -e, utriculatus -a -um*   with bladders or utricles bladder-like

*-utus -a -um*   -having

*uva-crispa*   curly-bunch (derivation doubtful)

*uvaria*   from an old generic name, clustered, like a bunch of grapes

*uva-ursi*   bear's-berry (Latin equivalent of the name *Arctostaphylos*)

*uva-vulpis*   fox's-berry

*uvifer -era -erum*   grape-bearing

*uviformis -is -e*   in a clustered mass

*Uvularia*   Palate (either from the hanging flowers or the fruits)


*Vaccaria*   Cow-fodder (an old generic name from 'vacca' a cow)

*vaccini-*   bilberry-like, resembling *Vaccinium*

*Vaccinium*   a name of great antiquity (like, *Hyacinthus*) with no clear meaning

*vaccinus -a -um*   the colour of a red cow, of cows

*vacillans*   variable

*vagans*   of wide distribution, wandering

*vagensis -is -e*   from the River Wye (Vaga)

*vaginans, vaginatus -a -um*   having a sheath, sheathed (as the stems of grasses by the leaf-sheaths)

*vagus -a -um*   uncertain, wandering

*vaillantii, Valantia*   for S. Vaillant (Valantius), French botanist (1669–1722)

*valdivianus -a -um, valdiviensis -is -e*   from Valdivia, Chile

*valentinus -a -um*   from Valencia, Spain

*Valeriana*   Health (from a medieval name for its medicinal use)

*Valerianella*   diminutive of the name *Valeriana*

*valesiacus -a -um*   from Valois, north-east France

*validus -a -um*   well-developed, strong

*vallesiacus -a -um, vallesianus -a -um*   from Vallais (Wallis), Switzerland

*Vallisneria*   for Vallisneri de Vallisnera of Padua (1661–1730)

*Vallota*   for A. Vallot, French botanist (1594–1671)

*valverdensis -is -e*   from Valverde, Hierro, Canary Isles

*Vanda*   from the Sanscrit name

*Vanilla*   Little-sheath (from the Spanish name describing the fruit)

*vari-, varii-, varius -a -um*   differing, changing, diverse, varying

*variabilis -is -e, varians*   variable

*variatus -a -um*   several, various

*varicosus-a -um*   with dilated veins or filaments, varicose

*variegatus -a -um*   irregularly coloured, variegated

*variifolius -a -um*   variable-leaved

*variolatus -a -um*  pock-marked, pitted

*vastus -a -um*  vary large, vast

*vectensis -is -e*  from the Isle of Wight (Vectis)

*vegetus -a -um*  growing strongly or quickly, vigorous

*Veitchia, veitchii*  for the Veitch nurserymen of Chelsea

*velaris -is -e*  veiling

*velox*  rapid-growing

*Veltheimia*  for A. F. G. von Veltheim, German patron of botany (1741–1808)

*veluti-, velutinus -a -um*  with a soft silky covering, velvety

*venator*  hunter (the flowers are 'hunting-pink')

*venenatus -a -um*  poisonous

*venetus -a -um*  of Venice, Venetian

*venosus -a -um*  conspicuously veined

*Ventenata*  for E. P. Ventenat, French botanist (1757–1808)

*ventricosus -a -um*  bellied out below, distended, expanded, ventricose

*venulosus -a -um*  finely-veined

*venustus -a -um*  graceful, beautiful, charming

*Veratrum*  the Latin name

*verbanensis -is -e*  from the area of Lake Maggiore

*verbasci-*  Mullein-like, resembling *Verbascum*

*Verbascum*  a name in Pliny

*Verbena*  the Latin name for the leafy twigs used in wreaths for ritual use and medicine

*verbenaca, verbeni-*  from a name in Pliny, vervain-like

*Verbesina*  *Verbena*-like (resembles some species)

*verecundus -a -um*  modest

*veris*  of spring (flowering time), genuine, true, standard

*vermi-, vermicularis -is -e, vermiculatus -a -um*  worm-like

*vernalis -is -e, vernus -a -um*  of spring (flowering time), vernal

*vernicifer -era -erum*  producing varnish

*vernicifluus -a -um*  from which flows a varnish

*vernicosus -a -um*  glossy, varnished

*vernix*  varnish

*Vernonia*  for Wm. Vernon, English botanist (c. 1680–1711)

*Veronica*  for St Veronica who wiped the sweat from Christ's face

*verrucosus -a -um*  with a warty surface, warted, verrucose

*versicolor*  varying or changeable in colour

*verticillaris -is -e*  having whorls (several leaves of flowers all arising at the same level on the stem), verticillate

*verticillaster*  with whorls of flowers

*verticillatus -a -um*   arranged in whorls, verticillate

*veruculosus -a -um*   somewhat warty

*verus -a -um*   true, genuine

*vescus -a -um*   small, feeble, fine, edible

*vesicarius -a -um*   inflated, bladder-like

*vesicatorius -a -um*   blistering

*vespertilis -is -e*   bat-like (with two large lobes)

*vespertinus -a -um*   of the evening (evening-flowering)

*vestalis -is -e*   white

*vestitus -a -um*   covered, clothed (with hairs)

*Vetiveria*   Latinized English version of southern Indian name for khus-khus grass

*Vetrix*   Osier

*vexans*   annoying, wounding

*vexillaris -is -e*   with a standard (as the large 'sail' petal of a pea-flower)

*vialis -is -e, viarum*   ruderal, of the wayside

*viatoris*   of the road-ways, of travellers

*Viburnum*   the Latin name for the wayfaring tree

*Vicia*   the Latin name for a vetch

*viciae-, vicii-*   vetch-like-, resembling *Vicia*

*vicinus -a -um*   neighbouring

*victorialis -is -e*   victorious (protecting)

*Vigna*   for D. Vigna, Italian scientist

*vilis -is -e*   common, of little value

*villicaulis -is -e*   with a shaggy stem

*villipes*   with a long-haired stalk

*villosulus -a -um*   slightly hairy, finely villous

*villosus -a -um*   with long soft hairs, shaggy, villous

*vilmorinianus -a -um, vilmorinii*   for the French nurserymen Vilmorin-Andrieux

*viminalis -is -e, vimineus -a -um*   with long slender shoots suitable for wicker or basketwork, of osiers, osier-like

*vinaceus -a -um*   of the vine, wine-coloured

*Vinca*   Binder (the Latin name refers to its use in wreaths)

*Vincetoxicum*   Poison-beater (its supposed antidotal property to snakebite)

*vincoides*   periwinkle-like, resembling *Vinca*

*vinculans*   binding, fettering

*vindobonensis -is -e*   from Vienna (Vindobona), Viennese

*vinealis -is -e*   of vines and vineyards, growing in vineyards

*vinicolor*   wine-red

*vinifer -era -erum*   wine-bearing

*vinosus -a -um*   wine-red, wine-like

*Viola*   the Latin name applied to several fragrant plants

*violaceus -a -um*   violet-coloured

*violescens*   turning violet

*viperatus -a -um*   viper-like (markings)

*viperinus -a -um*   snake's, serpent's

*virens*   green

*virescens*   light-green, turning green

*virgatus -a -um*   with straight slender twigs, twiggy

*Virgaurea*   Rod-of-gold

*virginalis -is -e, virgineus -a -um*   maidenly, purest white, virginal

*virginianus -a -um, virginiensis -is -e*   from Virginia, USA, Virginian

*virginicus -a -um*   from the Virgin Islands, Virginian

*virgulatorum*   of thickets

*viridescens*   becoming green, turning green

*viridi-, viridis -is -e*   fresh-green, youthful

*viridior*   more green, greener

*viridulus -a -um*   greenish

*virmiculatus -a -um*   vermillion

*virosus -a -um*   slimy, rank, poisonous

*Viscaria*   Bird-lime (the sticky stems of German catchfly)

*viscatus -a -um*   clammy

*viscidi-, viscidus -a -um, viscosus -a -um*   sticky, clammy, viscid

*Viscum*   the Latin name for bird-lime or mistletoe

*vitaceus -a -um*   vine-like, resembling *Vitis*

*vitalba*   vine-of-white (the appearance of *Clematis* in the fruiting state)

*vitellinus -a -um*   dull reddish-yellow, egg-yolk-yellow

*Vitex*   a name used in Pliny

*viti-*   vine-like, resembling *Vitis*

*viticella*   like a small vine

*vitiensis -is -e*   from the Fijian Islands (Viti Levu)

*Vitis*   the Latin name for the grapevine

*vitis-idaea*   vine of Mt Ida or Idaea, Greece

*vitreus -a -um*   glassy

*Vittaria*   Ribbon

*vittatus -a -eum*   striped lengthwise

*vittiformis -is -e*   band-like

*vittiger -era -erum*   bearing lengthwise stripes

*vivax*   long-lived (flowering for a long time)

*viviparus -a -um*   producing plantlets (often in place of flowers or as precocious germination on the parent plant), viviparous

*Voandzeia*   from the Madagascan name for the underground bean

*Vogelia, vogelii*   for J. R. T. Vogel of the Niger expedition, (1812–1841)

*volgaricus -a -um*   from the river Volga, Russia

*volubilis -is -e*   twining, entwining, enveloping

*volutus -a -um*   with rolled leaves, rolled

*vomitorius -a -um*   causing regurgitation, emetic

*Vriesia*   for W. H. deVries, Dutch botanist (1806–1862)

*vulcanicus -a -um*   fiery, of volcanoes or volcanic soils

*vulgaris -is -e, vulgatus -a -um*   usual, common, vulgar

*vulnerarius -a -um, vulnerum*   of wounds (wound-healing property)

*Vulpia*   for J. S. Vulpius, German botanist (1760–1846)

*vulpinus -a -um*   foxy, of the fox (coloration, but also implies inferiority)

*vulvaria*   cleft, with two ridges, of the vulva (the smell of *Chenopodium vulvaria*)


*Wahlenbergia*   for G. Wahlenberg, Swedish botanist (1780–1851)

*wardii*   for Frank Kingdon-Ward, collector of East Asian plants

*warleyensis -is -e*   of Warley Place, Essex (home of Miss Ellen Ann Willmott)

*Washingtonia*   for George Washington (1732–1799)

*watermaliensis -is -e*   from Watermal, Belgium

*Watsonia, watsonianus -a -um, watsonium*   for Sir Wm. Watson, scientist (1715–1787)

*watsonii*   for Wm. Watson, Curator of Royal Botanic Gardens, Kew (1858–1925)

*wichuraianus -a -um*   for Max E. Wichura, German botanist (1817–1866)

*Widdringtonia*   for Capt. Widdrington, botanist explorer

*willmottiae, willmottianus -a -um*   for Miss Ellen Ann Willmott, plant introducer (1858–1934)

*wilsonii*   for several including Dr E. H. Wilson who collected in the east for Veitch and the Arnold Arboretum and G. F. Wilson of Wisley

*Wolffia*   for J. F. Wolff, German doctor (1778–1806)

*wolgaricus -a -um*   from the region of the River Volga, Russia

*Woodsia*   for Joseph Woods, English botanist (1776–1864)

*xalapensis -is -e*   from Xalapa, Mexico
*xanth-, xanthi-, xantho-, xanthinus -a -um*   yellow-, yellow
*Xanthium*   Dioscorides' name for cocklebur, from which a
   yellow hair dye was made
*xanthospilus -a -um*   yellow-spotted
*xanthostephanus -a -um*   with a yellow crown
*xanthoxylum(on)*   yellow-wooded
*xero-*   dry-
*xerophilus -a -um*   drought-loving, living in dry places
*Xiphium*   Sword (the Greek name for a *Gladiolus*)
*xiphochilus -a -um*   with a sword-shaped lip
*xiphoides*   sword-like, shaped like a sword
*xiphophyllus -a -um*   with sword-shaped leaves
*xylo-*   wood-, woody-
*xylocanthus -a -um*   woody-thorned
*xylosteum*   hard-wooded (wood-bone)
*Xyris*   Greek name for a plant with razor-like leaves
*Xysmalobium*   Fragmented-lobes (of the corona)

*yakusimanus -a -um*   from Yakushima, Japan
*yedoensis -is -e*   from Tokyo (Yedo), Japan
*yemensis -is -e*   from the Yemen, Arabia
*yosemitensis -is -e*   from the Yosemite Valley, California, USA
*Yucca*   from a Carib name formerly applied to cassava
*yuccifolius -a -um*   with *Yucca*-like leaves
*yunnanensis -is -e*   fromYunnan, China

*za-*   much-, many-, very-
*zaleucus -a -um*   very white
*zalil*   from an Afghan name for a *Delphinium*
*Zamia*   a name in Pliny refers to the possession of cones
*zamii-*   resembling *Zamia*
*Zannichelia*   for Zannichelli, Italian botanist
*Zantedeschia*   for G. Zantedeschi, Italian botanist (1773–1846)
*zanzibarensis -is -e, zanzibaricus -a -um*   from Zanzibar, East
   Africa
*zapota*   a South American name for the chicle tree, *Sapodilla*
*Zea*   from the Greek name for another cereal
*zebrinus -a -um*   from the Portuguese, meaning striped with
   different colours, zebra-striped
*Zelkova*   from the Caucasian name

*Zephyranthes*   West-wind-flower

*zephyrius -a -um*   western, flowering or fruiting during the monsoon season (for Indonesian plants)

*Zerna*   a Greek name (for the *Cyperus*-like spikelets)

*zetlandicus -a -um*   from the Shetland Isles

*zeylanicus -a -um*   from Sri Lanka (Ceylon), Singhalese

*zibethinus -a -um*   of the civet (the fruits of *Durio zibethina* are used to trap the Asiatic civet *Vivera zibetha*)

*Zingiber*   from a pre-Greek name, possibly from India

*Zizania*   an ancient Greek name for a wild plant

*zizanioides*   resembling *Zizania* (Canadian wild rice)

*zonalis -is -e, zonatus -a -um*   girdled with distinct bands or concentric zones

*zooctonus -a -um*   poisonous

*zoster-*   girdle-

*Zostera*   Ribbon (Theophrastus' name for a marine plant)

*zygis*   yoke-like (paired flowers)

*zygo-*   paired-, balanced-, yoked-

*Zygocactus*   Jointed-stem-cactus

*zygomeris -is -e*   with twinned parts

*zygomorphus -a -um*   bilateral, of balanced form

*Zygophyllum*   Yoked-leaves (some have conspicuously paired leaves)

# Bibliography

Adanson, M. (1763–64). *Familles des Plantes*. Paris.

Bailey, L. H. (1949). *Manual of Cultivated Plants*. Macmillan, New York.

Bateson, W. (1909). *Mendel's Principles of Heredity*. Cambridge.

Bauhin, C. (1620). *Prodromus Theatri Botanici*. Frankfurt.

Bauhin, C. (1623). *Pinax Theatri Botanici*. Basel.

Boerhaave, H. (1710 & 1720). *Index Plantarum*...Leiden.

Brickell, C. D. *et al.* (1980). International Code of Nomenclature for Cultivated Plants. In *Regnum Vegetabile*, **104**, Deventer.

Britten, J. & Holland, R. (1886). *A Dictionary of English Plant Names*. The English Dialect Society, London.

Brunfels, O. (1530–36). *Herbarium Vivae Eicones*...Strasbourg.

Caesalpino, A. (1583). *De Plantis Libri xvi*. Florence.

Camp, W. H., Rickett, H. W. & Weatherby, C. A. (eds) (1947). Rochester Code. *Brittonia* **6(1)**, 1–120. Chronica Botanica, Mass., USA.

Candolle, A. de (1867). *Lois de la Nomenclature Botanique*. H. Georg, Paris.

Candolle, A. P. de (1813). *Théorie Élémentaire de la Botanique*. Paris.

Chittenden, F. J. (ed.) (1951). *Royal Horticultural Society Dictionary of Gardening*, vols. 1–4, and *Supplements* (1956 and (ed. P. M. Synge) 1969). Oxford University Press, Oxford.

Cordus, V. (1561–63). *Annotationes in Pedacii Dioscorides*. Strasbourg.

Correns, C. (1900). G. Mendel's Regel über das Verhalten der Nachkommenschaft der Rassenbastarde. *Berichte*, **18**, 158.

Cube, J. von (1485). *German Herbarius*. Mainz.

Darwin, C. G. (1859). *The Origin of Species by Means of Natural Selection*. J. Murray, London.

Dioscorides, P. (1934). *De Materia Medica*. John Goodyer

translation of 1655 (ed. R. T. Gunther), Oxford University Press, Oxford.

Dodoens, R. (1583). *Stirpium Historiae Pemptades*. Antwerp.

l'Ecluse, C. (1583). *Stirpium Nomenclator Pannonicus*. Német-Hjvár.

Farr, E. R. *et al.* (1979–86). Index Nominum Genericorum and Supplement 1. In *Regnum Vegetabile*, **100, 101, 102**, and **113**, The Hague.

Fernald, M. L. (1950). *Gray's Manual of Botany*, The American Book Company, New York.

Fuchs, L. (1942). *De Historia Stirpium*...Basel.

Gilbert-Carter, H. (1964). *Glossary of the British Flora*. 3rd edn. Cambridge University Press, Cambridge.

Green, M. L. (1927). The history of plant nomenclature. *Kew Bulletin*, **403–15**.

Grew, N. (1682) *The Anatomy of Plants*. London.

Grigson, G. (1975) *An Englishman's Flora*. Hart Davis, St Albans.

International Orchid Commission (1985). *Handbook on Orchid Nomenclature and Registration*. 3rd edn. The International Orchid Commission, London.

Ivimey-Cook, R. B. (1974). *Succulents: A Glossary of Terms and Descriptions*. The National Cactus and Succulent Society, Oxford.

Jackson, B. D. (1960). *A Glossary of Botanical Terms*. 4th edn. Duckworth, London.

Jeffrey, C. (1978). *Biological Nomenclature*. 2nd edn. Edward Arnold, London.

Johnson, A. T. & Smith, H. A. (1958). *Plant Names Simplified*. Feltham.

Jung, J. (1747). Doxoscopiae (1662); Isagoge phytosocpica (1679). In *Opuscula Botanica-physica*. Coburg.

Jussieu, A. L. (1789). *de Genera Plantarum*. Paris.

Linnaeus, C. (1735). *Systema Naturae*. Leiden.

Linnaeus, C. (1738). *Classes Plantarum*. Leiden.

Linnaeus, C. (1751). *Philosophia Botanica*. Stockholm/Amsterdam.

Linnaeus, C. (1753). *Species Plantarum*. Stockholm.

Linnaeus, C. (1754). *Genera Plantarum*. 5th edn. Stockholm.

Linnaeus, C. (1759). *Systema Naturae*. 10th edn. Stockholm.

Linnaeus, C. (1762–3). *Species Planarum*. 2nd edn. Stockholm.

Linnaeus, C. (1764). *Genera Plantarum*. 6th edn. Stockholm.

MacLeod, R. D. (1952). *Key to the Names of British Plants*. Pitman & Sons, London.

Magnus, Albertus (1478) *Liber aggregationis seu liber secretorum Alberti magni de virtutibus herbarum*. Johann de Anunciata de Augusta.

Malpighi, M. (1671)...*Anatome Plantarum*...London (1675–79)

Mendel, G. J. (1866). *Versuche über Planzenhybriden*. Brno.

Mentzel, C. M. (1682). *Index Nominum Plantarum Multilinguis (Universalis)*. Berlin.

Morison, R. (1672). *Plantarum Umbelliferum Distributio Nova*...Oxford.

Morison, R. (1680). *Plantarum Historia Universalis*...Oxford.

l'Obel, M. (1576). *Plantarum seu Stirpium Historia*...Antwerp.

Paracelsus (Bombast von Hohenheim) (1570). *Dispensatory and Chirurgery...Faithfully Englished by W. D. London*, 1656.

Parkinson, J. (1629). *Paradisi in Sole Paradisus Terrestris*. H. Lownes & R. Young, London (Reprinted by Methuen, London, 1904).

Pliny Gaius secundus (AD23–79) Thirty-seven books of *Historia Naturalis*.

Plowden, C. C. (1970). *A Manual of Plant Names*. George Allen & Unwin, London.

Porta Giambatista Della (Johannes Baptista) (1588). *Phytognomica*. Naples.

Prior, R. C. A. (1879). *On the Popular Names of British plants*. 3rd edn. London.

Rauh, W. (1979). *Bromeliads*. English translation by P. Temple, Blandford Press, Dorset.

Ray, J. (1682). *Methodus Plantarum*. London.

Ray, J. (1686–1704). *Historia Plantarum*. London.

Rivinus, A. Q. (1690). *Introductio Generalis in Rem Herbariam*. Leipzig.

Schultes, R. E. & Pease, A. D. (1963). *Generic Names of Orchids: Their Origin and Meaning*. Academic Press, London.

Smith, A. W. (1972). *A Gardener's Dictionary of Plant Names*. (Revised and enlarged by W. T. Stearn) Cassell, London.

Sprague, T. A. (1950). The evolution of botanical taxonomy from Theophrastus to Linnaeus. In *Lectures on the Development of Taxonomy*. Linnean Society of London.

Stafleu, F. A. *et al.* (eds.) (1983). International Code of Botanical Nomenclature. In *Regnum Vegetabile*. **111**, Utrecht.

Stafleu, F. A. & Cowan, R. S. (1976–). Taxonomic Literature. 2nd edn. In *Regnum Vegetabile*, **94** (1976), **98** (1979), **105** (1981), **110** (1983), **112** (1985), **115** (1986). Utrecht.

Stearn, W. T. (1983). *Botanical Latin*. David & Charles, Newton Abbot.

Styles, B. T. (ed.) (1986). *Infraspecific Classification of Wild and Cultivated Plants*. Special Vol. 29. The Systematics Association, Oxford.

Sutton, W. S. (1902). On the morphology of the chromosome group in *Brachystola magna*. *Bio. Bull.* **4**, 24–39.

Theophrastus (1483). *De Causis plantarum, lib VI*. Bartholomaeum Confalonerium de Salodio.

Tournefort, J. P. de (1694). *Elemens de Botanique*. Paris.

Tournefort, J. P. de (1700). *Institutiones Rei Herbariae*. Paris.

Tschermak, E. (1900). Uber Künstliche Kreuzung bei *Pisum sativum*. *Biologisches Zentralblatt*, **20**, 593–5.

Turner, W. (1965). Libellus de Re Herbaria (1538); The Names of Herbes (1548). *The Ray Society*, **145**, London.

Turner, W. (1551–68). *A New Herbal*. London & Cologne.

U.P.O.V. (1985). *International Convention for the Protection of New Varieties of plants*. Texts of 1961, 1972, 1978. U.P.O.V. publication **293E**. Geneva.

deVries, H. (1900). Sur la loi de disjonction des hybrides. *Comptes Rendues…*, **130**, 845–7. Paris. (Das Spaltungsgesetz der Bastarde; Vorläufige Mitteilung. *Ber. Dtsch. Bot. Ges.* **18**, 83–90.)

Willis, J. C. (1955). *A Dictionary of Flowering Plants and Ferns*. 6th edn. Cambridge University Press, Cambridge.

Wilmott, A. J. (1950). Systematic botany from Linnaeus to Darwin. In *Lectures on the development of taxonomy*. Linnean Society of London.

Zimmer, G. F. (1949). *A Popular Dictionary of Botanical Names and Terms*. Routledge & Kegan-Paul, London.